Environmental Neurotoxicology

Committee on Neurotoxicology and Models for Assessing Risk

Board on Environmental Studies and Toxicology

Commission on Life Sciences

National Research Council

WITHDRAWN

NATIONAL ACADEMY PRESS
Washington, D.C. 1992

COLORADO COLLEGE LIBRARY
COLORADO SPRINGS
COLORADO

NATIONAL ACADEMY PRESS 2101 Constitution Ave., N.W.   Washington, D.C. 20418

NOTICE: The project that is the subject of this report was approved by the Governing Board of the National Research Council, whose members are drawn from the councils of the National Academy of Sciences, the National Academy of Engineering, and the Institute of Medicine. The members of the committee responsible for the report were chosen for their special competencies and with regard for appropriate balance.

This report has been reviewed by a group other than the authors according to procedures approved by a Report Review Committee consisting of members of the National Academy of Sciences, the National Academy of Engineering, and the Institute of Medicine.

The National Academy of Sciences is a private, nonprofit, self-perpetuating society of distinguished scholars engaged in scientific and engineering research, dedicated to the furtherance of science and technology and to their use for the general welfare. Upon the authority of the charter granted to it by the Congress in 1863, the Academy has a mandate that requires it to advise the federal government on scientific and technical matters. Dr. Frank Press is president of the National Academy of Sciences.

The National Academy of Engineering was established in 1964, under the charter of the National Academy of Sciences, as a parallel organization of outstanding engineers. It is autonomous in its administration and in the selection of its members, sharing with the National Academy of Sciences the responsibility for advising the federal government. The National Academy of Engineering also sponsors engineering programs aimed at meeting national needs, encourages education and research, and recognizes the superior achievements of engineers. Dr. Robert M. White is president of the National Academy of Engineering.

The Institute of Medicine was established in 1970 by the National Academy of Sciences to secure the services of eminent members of appropriate professions in the examination of policy matters pertaining to the health of the public. The Institute acts under the responsibility given to the National Academy of Sciences by its congressional charter to be an adviser to the federal government and, upon its own initiative, to identify issues of medical care, research, and education. Dr. Kenneth Shine is president of the Institute of Medicine.

Environmental neurotoxicology / Committee on Neurotoxicology and Models for Assessing Risk, Commission on Life Sciences, National Research Council
     p.   cm.
    Includes bibliographical references and index.
    ISBN 0-309-04531-2
    1. Neurotoxicology. 2. Environmental monitoring. 3. Health risk assessment. I. National Research Council (U.S.). Committee on Neurotoxicology and Models for Assessing Risk.
    [DNLM: 1. Environmental Exposure. 2. Nervous System—drug effects. WL 100 E61]
RC347.5.E58   1991
615.9—dc20
DNLM/DLC
for Library of Congress

91-43537
CIP

Additional copies of this report are available from the National Academy Press, 2101 Constitution Avenue, N.W., Washington, D.C. 20418

S538

Printed in the United States of America

First Printing, January 1992
Second Printing, August 1992

2C
347,5
ES8
991

Committee on Neurotoxicology and Models for Assessing Risk

PHILIP J. LANDRIGAN, *Chairman*, Mt. Sinai School of Medicine, New York
DOYLE G. GRAHAM, *Vice Chairman*, Duke University Medical Center, Durham, North Carolina
W. KENT ANGER, Oregon Health Sciences University, Portland
JEFFERY BARKER, National Institutes of Health, Bethesda, Maryland
TERRI DAMSTRA, National Institute of Environmental Health Sciences, Research Triangle Park, North Carolina
DALE HATTIS, Massachusetts Institute of Technology, Cambridge
WILLIAM LANGSTON, California Parkinson Foundation, San Jose
HERBERT E. LOWNDES, Rutgers University, Piscataway, New Jersey
JOE MARWAH, National Institutes of Health, Bethesda, Maryland
PIERRE MORELL, University of North Carolina, Chapel Hill
TOSHIO NARAHASHI, Northwestern University Medical School, Chicago
PHILLIP P. NELSON, National Institutes of Health, Bethesda, Maryland
LAWRENCE W. REITER, U.S. Environmental Protection Agency, Research Triangle Park, North Carolina
PATRICIA RODIER, University of Rochester, Rochester, New York
JOSEPH RODRICKS, Environ Corporation, Arlington, Virginia
ELLEN K. SILBERGELD, University of Maryland at Baltimore, Baltimore
PETER S. SPENCER, Oregon Health Sciences University, Portland
BERNARD WEISS, University of Rochester, Rochester, New York
RONALD WYZGA, Electric Power Research Institute, Palo Alto, California
DONALD MATTISON, Liaison, Board on Environmental Studies and Toxicology
JOHN L. EMMERSON, Liaison, Board on Environmental Studies and Toxicology

Staff

RICHARD D. THOMAS, Program Director
KATHLEEN R. STRATTON, Project Director
MARY B. PAXTON, Senior Staff Officer
MARVIN A. SCHNEIDERMAN, Senior Staff Scientist
ANDREW M. POPE, Senior Staff Officer
NORMAN GROSSBLATT, Editor
ANNE M. SPRAGUE, Information Specialist
GARY J. BENNETT, Technical Adviser
IAN C.T. NISBET, Technical Adviser
HUGH TILSON, Technical Adviser
LINDA V. LEONARD, Senior Project Assistant

Sponsor

Agency for Toxic Substances and Disease Registry, U.S. Public Health Service

Board on Environmental Studies and Toxicology

PAUL G. RISSER *(Chairman)*, University of New Mexico, Albuquerque
GILBERT S. OMENN *(Immediate Past Chairman)*, University of Washington, Seattle
FREDERICK R. ANDERSON, Washington School of Law, American University
JOHN C. BAILAR, III, McGill University School of Medicine, Montreal
LAWRENCE W. BARNTHOUSE, Oak Ridge National Laboratory, Oak Ridge
GARRY D. BREWER, Yale University, New Haven
EDWIN H. CLARK, Department of Natural Resources & Environmental Control, State of Delaware, Dover
YORAM COHEN, University of California, Los Angeles
JOHN L. EMMERSON, Lilly Research Laboratories, Greenfield, Indiana
ROBERT L. HARNESS, Monsanto Agricultural Company, St. Louis
ALFRED G. KNUDSON, Fox Chase Cancer Center, Philadelphia
GENE E. LIKENS, The New York Botanical Garden, Millbrook
PAUL J. LIOY, UMDNJ-Robert Wood Johnson Medical School, Piscataway, New Jersey
JANE LUBCHENCO, Oregon State University, Corvallis
DONALD MATTISON, University of Pittsburgh, Pittsburgh
GORDON ORIANS, University of Washington, Seattle
NATHANIEL REED, Hobe Sound, Florida
MARGARET M. SEMINARIO, AFL/CIO, Washington, DC
I. GLENN SIPES, University of Arizona, Tucson
WALTER J. WEBER, JR., University of Michigan, Ann Arbor

Staff

JAMES J. REISA, Director
DAVID J. POLICANSKY, Associate Director and Program Director for Applied Ecology and Natural Resources
RICHARD D. THOMAS, Associate Director and Program Director for Human Toxicology and Risk Assessment
LEE R. PAULSON, Program Director for Information Systems and Statistics
RAYMOND A. WASSEL, Program Director for Environmental Sciences and Engineering

Commission on Life Sciences

BRUCE M. ALBERTS (*Chairman*), University of California, San Francisco
BRUCE N. AMES, University of California, Berkeley
J. MICHAEL BISHOP, Hooper Research Foundation, University of California Medical Center, San Francisco
MICHAEL T. CLEGG, University of California, Riverside
GLENN A. CROSBY, Washington State University, Pullman
LEROY E. HOOD, California Institute of Technology, Pasadena
DONALD F. HORNIG, Harvard School of Public Health, Boston
MARIAN E. KOSHLAND, University of California, Berkeley
RICHARD E. LENSKI, University of California, Irvine
STEVEN P. PAKES, Southwestern Medical School, University of Texas, Dallas
EMIL A. PFITZER, Hoffman-LaRoche, Inc., Nutley, New Jersey
THOMAS D. POLLARD, Johns Hopkins Medical School, Baltimore
JOSEPH E. RALL, National Institutes of Health, Bethesda, Maryland
RICHARD D. REMINGTON, University of Iowa, Iowa City
PAUL G. RISSER, University of New Mexico, Albuquerque
HAROLD M. SCHMECK, JR., Armonk, New York
RICHARD B. SETLOW, Brookhaven National Laboratory, Upton, New York
CARLA J. SHATZ, Stanford University School of Medicine, Stanford
TORSTEN N. WIESEL, Rockefeller University, New York, NY

JOHN E. BURRIS, Executive Director

The National Research Council was organized by the National Academy of Sciences in 1916 to associate the broad community of science and technology with the Academy's purposes of furthering knowledge and advising the federal government. Functioning in accordance with general policies determined by the Academy, the Council has become the principal operating agency of both the National Academy of Sciences and the National Academy of Engineering in providing services to the government, the public, and the scientific and engineering communities. The Council is administered jointly by both Academies and the Institute of Medicine. Dr. Frank Press and Dr. Robert M. White are chairman and vice chairman, respectively, of the National Research Council.

The project was supported by the Environmental Protection Agency; the National Institute of Environmental Health Sciences; and the Comprehensive Environmental Response, Compensation, and Liability Act Trust Fund through cooperative agreement with the Agency for Toxic Substances and Disease Registry, U.S. Public Health Service, Department of Health and Human Services.

Preface

There is convincing evidence that chemicals in the environment can alter the function of the nervous system. The number of people afflicted with neurotoxic disease can only be estimated, because the number of neurotoxic substances is unknown but probably plentiful, and the effects on the nervous system are many and varied. Despite increasing attention to neurotoxicity in recent years, much work still needs to be done.

The Board on Environmental Studies and Toxicology of the National Research Council's Commission on Life Sciences convened the Committee on Neurotoxicology and Models for Assessing Risk. Support was provided by the Agency for Toxic Substances and Disease Registry of the U.S. Public Health Service. The charge to the committee was to 1) assess the biologic bases of neurotoxicity with regard to establishing underlying principles relevant to risk assessment and extrapolation across species; 2) review existing models and indicators of neurotoxic action and disease, including structure-activity relationships, with respect to their efficacy in identifying neurotoxicants from environmental, occupational, and other potential sources; and 3) develop critical hypotheses for future research in neurotoxicology, particularly research that will lead to models for assessing the risks of neurotoxic disease. Committee members represented the breadth of disciplines involved in environmental neurotoxicology. Their expertise served the committee well as it endeavored to meet its charge.

The committee met seven times over 2 years. It focused on the magnitude of the problem, the use of biologic markers, neurotoxicity testing, surveillance and epidemiology, and risk assessment. This report presents the views of the committee members, and the conclusions and recommendations reflect the committee's deliberations. The committee concluded that research is needed to improve understanding of the mechanisms of neurotoxic action and the extent of neurotoxic disease. The committee encouraged the development of a coherent, tiered testing strategy and the improvement of surveillance programs.

The committee acknowledges the tireless efforts of those without whom the report would never have been completed. The committee thanks Hugh Tilson for his expertise and insight. The committee also acknowledges NRC staff for their work in organizing and managing this undertaking. Devra Davis was instrumental in conceptualizing the study initially. James Reisa, director of the Board on Environmental Studies and Toxicology, provided much welcomed guidance and encouragement. In addition to his contributions as program director, Richard

Thomas served as interim project director and gave invaluable support to the committee and staff. The committee thanks senior staff officers Andrew Pope and Mary Paxton for their work in the formative stages of the committee process and in helping with early drafts of the report, and Kathleen Stratton for seeing the report through its final stages. The committee recognizes the tireless efforts of Anne Sprague of the Toxicology Information Center and Linda Leonard, project assistant. Norman Grossblatt and Lee Paulson served as editors.

On behalf of the committee, I thank all who assisted in completing this report.

Philip J. Landrigan, *Chairman*
Committee on Neurotoxicology and
Models for and Assessing Risk

Contents

EXECUTIVE SUMMARY   1
  Magnitude of the Problem, 1
  Neurotoxicity Testing, 2
  Surveillance and Epidemiologic Studies in Environmental Neurotoxicology, 4
  Biologic Markers in Environmental Neurotoxicology, 5
  Neurotoxicity Risk Assessment, 5

1 INTRODUCTION: DEFINING THE PROBLEM OF NEUROTOXICITY   9
  Neurologic Responses to Environmental Toxicants, 9
  Magnitude of the Problem of Neurotoxicity, 17
  Detection and Control of Exposure to Neurotoxicants, 18
  Scope of this Report, 19

2 BIOLOGIC BASIS OF NEUROTOXICITY   21
  Cellular Anatomy and Physiology, 21
  General Aspects of Nervous System Structure and Function, 27
  Vulnerability of the Nervous System to Chemical Toxicants, 30
  Examples of Neurotoxic Mechanisms, 33
  MPTP, 39
  Summary, 40

3 BIOLOGIC MARKERS IN NEUROTOXICOLOGY   43
  Concepts and Definitions, 43
  Validation of Biologic Markers, 48
  Use of Biologic Markers in Risk Assessment, 50
  Summary, 51

4 TESTING FOR NEUROTOXICITY   53
  Approach to Neurotoxicity Testing, 54

Current Methods Based on Structure-Activity Relationships, 58
Current In Vitro Procedures, 59
Current In Vivo Procedures, 65
Current Regulatory Approaches, 86
Strategies for Improved Neurotoxicity Testing, 88

5 SURVEILLANCE TO PREVENT NEUROTOXICITY IN HUMANS   95
Neurobehavioral Test Batteries, 97
Current Exposure-Surveillance Efforts, 104
Current Disease-Surveillance Efforts, 105

6 RISK ASSESSMENT   111
Approaches to Risk Assessment for Neurotoxicity, 114
Curve-Fitting in Risk Assessment for Neurotoxicity, 116
Mechanistic Models for Risk Assessment for Neurotoxicity, 117
Summary, 120

7 CONCLUSIONS AND RECOMMENDATIONS   123

REFERENCES   129

INDEX   149

Tables and Figures

TABLES

1-1   Partial List of Neurotoxicants, 10
1-2   Human and Animal Neurobehavioral Effects Attributed to at Least 25 Chemicals, 11
1-3   Selected Major Neurotoxicity Events, 12
2-1   Nonneuronal (Glial) Cells of the Nervous System and Their Function, 22
2-2   Neuron Type Classified by Neurochemical Released for Synaptic Transmission, 27
3-1   Examples of Characteristics of Exogenous Agents, Organisims, or Targets That Influence Choice of Biologic Marker, 44
3-2   Selected Markers of Neurotoxicity in Nervous System, 45
4-1   In Vitro Neurobiologic Test Systems, 62
4-2   Markers for Assessing Neurotoxicity in In Vitro Systems, 63
4-3   Proposed Protocol for Developing and In Vitro Neurotoxicity Screening System, 65
4-4   Neurotoxic Effects of Representative Agents in Humans and Animals, 67
4-5   Examples of Behavioral Measures of Functional Neurotoxicity, 69
4-6   End Points That Might be Included in a Functional Observational Battery, 71
4-7   Areas of the Nervous System to be Used in Neuropathologic Evaluation, 81
4-8   Tissues of the Nervous System to be Used in Neuropathologic Evaluation, 82
4-9   Tests Used in NCTR Collaborative Study, 84
4-10   Proposed Components for Evaluating In Vitro Neurotoxicity Screening Tests, 92
5-1   Characteristics of Responses to Exposure to Some Neurotoxicants, 98
5-2   Components of Clinical Neurologic Examination, 100
5-3   Test Batteries, 102
6-1   Some Neurotoxicants That Act on Receptors, 118

FIGURES

2-1   Diagrammatic representation of neuronal structure, 23
2-2   Events in chemical synaptic transmission, 26
2-3   Structures of type I and type II pyrethroids, 34
2-4   Metabolism of hexane, 37
2-5   Diagram of MPTP toxicity, 39
3-1   Simplified classification of biologic markers, 46
4-1   Biologic markers in the stages between formation and degneration of neural circuits, 91
6-1   Effect of a shift in mean IQ score on the population distribution, 113
6-2   Estimated developmental scores at various ages for three blood-level concentrations, 115
6-3   Percentage of severe mental retardation among those exposed in utero by dose and gestational age in Hiroshima and Nagasaki, 119

Executive Summary

The recognition that exposure to chemicals can cause neurologic injury evolved from studies of acute illnesses in people exposed to high doses of environmental toxicants. These illnesses included encephalopathy in children who ate chips of lead-based paint; blindness in persons who consumed wood alcohol (methanol); and coma, convulsions, and respiratory paralysis after exposure to organophosphorus pesticides. Epidemics of neurotoxic diseases related to environmental exposures have occurred: blindness and ataxia caused by organic mercury in fish from Minamata Bay, Japan, and in fungicide-treated grain in Iraq; spinal-cord degeneration and peripheral neuropathy caused by tri-o-cresylphosphate (TOCP) in cooking oil in Morocco; and tremors, motor disturbance, and anxiety caused by the pesticide Kepone (chlordecone) in Hopewell, Virginia. In all, these epidemics affected thousands of people and established clearly that toxic chemicals in the environment[1] can cause neurologic and psychiatric illnesses.

In response to a request from the Agency for Toxic Substances and Disease Registry, the National Research Council convened the Committee on Neurotoxicology and Models for Assessing Risk in 1988. The committee, charged to review the biologic principles and mechanisms of neurotoxic action relevant to risk assessment, produced this report, which discusses the magnitude of the problem of neurotoxic effects, testing strategies, surveillance efforts, biologic markers, and risk assessment.

MAGNITUDE OF THE PROBLEM

Neurotoxicity caused by environmental toxicants results in a range of neurologic and

---

[1]"Environment" is defined broadly in this document to encompass a wide range of external factors that can cause injury, including diet, ethanol, tobacco, drugs, and occupational exposures, as well as toxic contaminants in what are ordinarily considered components of the ambient environment—air, water, and soil. "Neurotoxicity" is defined as the capacity of chemical, biologic, or physical agents to cause adverse functional or structural changes in the nervous system.

psychiatric disorders. Concern over the potential neurotoxic effects of chemical substances is greatest for agents that cause irreversible or progressive changes. In addition to immediate and progressively developing effects, there is increasing evidence that neurotoxic effects can occur after long latent periods. It is postulated that intervals as long as many decades can elapse between exposure to a chemical and the appearance of neurologic illness.

A major unanswered question—indeed, a central issue confronting neurotoxicology today—is whether the causal associations observed in epidemics of neurotoxic diseases reflect isolated events or are merely the most obvious examples of a widespread association between environmental chemicals and nervous system impairment. Concern about subclinical neurotoxicity has brought this issue to its current prominence. Subclinical toxicity refers to exposure-induced adverse effects that are too small to produce signs and symptoms evident in a standard clinical examination. Subclinical neurotoxic effects can include alteration of a wide spectrum of behaviors. Environmental chemicals associated with subclinical neurotoxicity include lead, organophosphorus pesticides, some chlorinated hydrocarbons, some solvent mixtures, and mercury. Those are chemicals to which many thousands of Americans are regularly exposed at work and to which even more are exposed in smaller doses in the general environment. Although often subtle, subclinical neurotoxic effects are not necessarily inconsequential; moreover, even subtle alterations can be irreversible. It has been hypothesized that an undefined fraction of chronic neurologic and psychiatric illness in the human population can be exacerbated or even caused by chronic, low-level exposure to environmental neurotoxicants.

The committee recommends that more accurate estimates of the extent of the problem of neurologic and psychiatric dysfunction attributable to chemical agents in the environment be developed.

The estimates must be based on a combination of clinical, epidemiologic, and toxicologic studies coupled with the techniques of quantitative risk assessment.

NEUROTOXICITY TESTING

The general reason to test substances for neurotoxicity is to identify neurotoxic potential before the occurrence of human exposure. The ultimate goal is to prevent human disease. About 70,000 chemicals are used in commerce, of which several hundred are known to be neurotoxicants. However, except for pharmaceuticals, less than 10% of the chemicals in commerce have been tested at all for neurotoxicity, and only a handful have been evaluated thoroughly. Furthermore, resources are not readily available to undertake across-the-board testing of all chemical substances already in commerce.

New strategies for neurotoxicologic assessment of environmental chemicals will need to include establishing testing priorities among chemicals for hazard identification (with emphases on new chemicals, chemicals considered likely to be hazardous, and chemicals to which large numbers of people are exposed). The new strategies will also require refining existing neurotoxicity test systems and developing sensitive, new testing systems. To contain the labor and resource requirements of the testing strategy, developing quicker and more economical approaches than are currently used must be encouraged, particularly for screening for potential neurotoxicants.

A new strategy for neurotoxicologic assessment will build on and extend currently available test systems. It will have a "tiered" structure—decisions to test chemicals at the higher tiers, as well as decisions concerning types of testing, will be guided by data from the initial, or screening, tier. The first tier, or screen, will be used for hazard identification. The results of the screen and a chemical's exposure pattern would guide further characterization of dose-response relationships (second tier) and mechanisms (third tier).

No existing validated system satisfies all the necessary requirements for a screening program to detect the neurotoxic potential of chemicals. The range of such a program should extend to the detection of neuro-developmental effects and effects on cognitive function and of neuroendocrine effects. No comprehensive effort has been made to determine the predictive ability of individual screening tests by examining the relationship between test results and data from long-term studies in animals or epidemiologic and clinical studies in humans.

The screening tier will consist of a set of tests to measure multiple end points—including chemical, structural, and functional changes—and a functional observational battery. Determination of the component tests of the first tier is the most crucial aspect of the three-tiered approach, because truly positive substances will not continue to later tiers unless detected at this point. The first tier is the heart of the screening aspect of neurotoxicity testing; the later tiers might produce data of great value in advancing understanding of neurotoxic processes and how the nervous system operates, but the first tier is the first line in preventing neurotoxic disease.

Much of the controversy over proper testing procedures arises from two intrinsically conflicting objectives—minimizing the incidence of false positives (substances incorrectly identified as hazardous) and minimizing the incidence of false negatives (substances incorrectly identified as nonhazardous). Too high an incidence of false positives wastes resources. However, a high incidence of false negatives is potentially dangerous. For the prevention of human disease, testing systems must be highly sensitive. Testing in the second and third tiers of future test systems should expose false positives; the abandonment of an occasional new chemical on the basis of what are actually false-positive screening results is the likely cost of this process.

The committee recommends that a rational, cost-effective neurotoxicity testing strategy be adopted.

It should allow an accurate and efficient progression from the results of hazard-identification studies (screening) to the selection and application of appropriate test methods for defining mechanisms of toxicity and for quantitative characterization of neurotoxic hazards.

A coherent testing strategy would incorporate a series of checks and balances. The results of the later phases of testing would provide the data necessary to evaluate and validate initial screening batteries and thereby help to identify tests that should be excluded from or incorporated into an efficient battery. Information generated by such a strategy would reveal which types of data are most useful for accurate, quantitative prediction of the risks to humans associated with exposure to similar chemical compounds.

Tests of neurotoxicity on chronically exposed animals might be carried out in conjunction with tests of other chronic effects. Such testing might identify toxic effects of substances not previously known to be neurotoxic. There is a particular lack of data on chronic and long-latency neurotoxic effects.

The committee recommends that for reasons of efficiency, integrative studies combining a variety of end points be explored in the development of the neurotoxicity testing strategy.

To maximize detection of toxicity, some toxicity studies encompassing the full life span of experimental animals should be encouraged.

Data are needed on the influence of dose, route of exposure, toxicokinetics, metabolism, and elimination on the effects of a given neurotoxicant, and data are needed on the existence of interspecies differences. More complete understanding of neurotoxic disease at a molecular level should also improve the ability to evaluate new chemicals on the basis of structure-activity relationships. Existing in vitro test methods should be exploited more efficiently than at present to identify and analyze the mechanisms of neurotoxic action at cellular levels. Development of a mechanistic understanding of neurotoxicity might facilitate the discovery of biologic markers of exposure to toxicants, as well as markers of early, subclinical neurotoxic effects.

The committee recommends that studies to define mechanisms of neurotoxicity in as much detail as possible be encouraged, as well as studies to identify hazards.

A program needs to be undertaken to test the relationship between in vitro and in vivo findings and between animal and human results for a set of well-defined substances.

SURVEILLANCE AND EPIDEMIOLOGIC STUDIES IN ENVIRONMENTAL NEUROTOXICOLOGY

Epidemiologic and clinical studies of populations exposed to potentially neurotoxic chemicals are needed to provide additional information on the human neurotoxic effects of environmental chemicals and to complement in vitro and animal screening studies. High-risk populations must be identified and monitored. Public-health surveillance systems to identify people who are potentially exposed to environmental neurotoxicants are not well developed, and there is little information on the background incidence and prevalence of the major neurologic diseases in the American population. People with diagnosed neurologic illnesses must be studied to identify possible environmental etiologies and to complement and extend the knowledge gained through in vivo and in vitro laboratory investigations.

Recognition of the neurotoxic effects of exposure to environmental chemicals through epidemiologic and clinical studies is made difficult by the enormous variety of the possible reactions of the nervous system to toxic insult. The changes are often subtle and subclinical, and months or years can elapse between exposure to a neurotoxicant and the appearance of dysfunction or disease. Few attempts have been made to explore the possible relationships between chemical exposures and chronic or progressive neurologic and behavioral disorders. Populations known to be exposed to potential neurotoxicants should be followed for long periods in prospective studies, and retrospective studies of people with neurologic illness must consider the possibility that exposures occurred many years previously. Epidemiologic studies will increasingly need to use biologic markers of exposure, of effects, and of susceptibility.

The committee recommends that exposure-surveillance systems cover a much broader range of chemicals and use improved monitoring techniques for long-term assessment.

Existing disease-surveillance systems, such as those of the Social Security Administration, the Department of Veterans Affairs, and the National Center for Health Statistics, should be modified to provide more useful data on the incidence and prevalence of chronic neurologic and psychologic disorders, some of which are likely to be of occupational and environmental origin. A broader range of neurologic disease end points should be covered by surveillance programs. Anecdotal reports of neurotoxicity in humans need to be pursued vigorously with clinical surveillance and followup. The incorporation in surveillance systems of the concept of sentinel health events (SHEs) specifically for neurotoxic illnesses should be encouraged.

Recognition of the possible environmental origin of neurologic and psychiatric disease is hampered by the inadequate training of most physicians and other health providers in occupational and environmental medicine. Greater uniformity in disease definition would improve identification of diseases of neurologic interest.

The committee recommends that improved disease reporting be supported by the dissemination of information on neurotoxic illnesses to physicians and other health professionals to increase their awareness of environmental neurotoxicity as a possible explanation of specific illnesses or sets of symptoms.

All physicians should be trained to take a thorough occupational-exposure history and to be aware of other possible sources of toxic exposure, such as hobbies and self-medication. Uniform clinical definitions of neurotoxic disorders are needed to provide a common basis for reporting by physicians. Standardized national reporting systems must be established for physicians to report outbreaks of suspected environmentally and occupationally caused neurologic and psychiatric disorders.

BIOLOGIC MARKERS IN ENVIRONMENTAL NEUROTOXICOLOGY

As recently defined by a previous NRC committee, biologic markers are measures of changes or variations in biologic systems or samples. It is useful to classify biologic markers into three types: markers of exposure, of effect, and of susceptibility. A biologic marker of exposure in an organism is a measurable presence of an exogenous substance, its metabolite, or the product of its interaction with some target molecule or cell. A biologic marker of effect is a measurable biochemical, physiologic, or other alteration within an organism that, depending on magnitude, can indicate potential or established disease. A biologic marker of susceptibility is an indicator of an inherent or acquired variation in an organism's ability to respond to the challenge of exposure to a specific substance.

Biologic markers provide a means to explore the relationship between exposures and effects. By virtue of the incorporation of more relevant information about critical events and molecular mechanisms, biologic markers contribute to a more complete scientific understanding of toxic injury. Biologic markers can be used to improve sensitivity, specificity, and predictive value of detection and quantification of adverse effects at low dose and early in exposure. Neuroscience and neurotoxicology can be enhanced by directing more research attention to quantitative questions that information on biologic markers helps to answer.

The committee recommends that putative biologic markers in animal species be evaluated and validated in in vivo and in vitro systems.

Biologic markers should be regularly incorporated into epidemiologic and clinical studies of neurologic disease, particularly prospective studies. The primary goal of the incorporation of biologic markers into such studies should be to validate their predictive accuracy and to test hypothesized quantitative relationships between specific markers related to causal pathways involving neurotoxic outcomes.

NEUROTOXICITY RISK ASSESSMENT

Risk-assessment techniques provide a means for estimating the risks to humans associated with exposure to toxic chemicals in the environment. The estimation of the risks most often involves extrapolation from high experimental doses used in animal tests to lower environmental doses. Numerous assumptions are typically made to bridge gaps in the available scientific data. Most risk-assessment procedures have focused on cancer as an end point, and techniques for assessing other types of risk are relatively undeveloped. Commonly used paradigms for risk assessment do not accurately or adequately model the risks associated with exposure to neurotoxicants. Virtually all neurotoxicologic risk assessment today is limited to qualitative hazard identification and to the early stages of hazard characterization; neither sufficient data nor adequate paradigms are available to permit quantitative evaluation of most neurotoxic risks.

Risk-assessment techniques that incorporate more quantitative information about dose-time-response relationships and mechanisms of neurotoxicity are under development. Advances in this regard will improve not only the assessment of risk, but also assessment of the benefits to be gained by reducing human exposure to specific neurotoxic agents. The construction of new models for neurotoxicologic risk assessment depends on the acquisition of new knowledge of the fundamental mechanisms of action of chemical toxicants on the human nervous system. The molecular and subcellular mechanisms by which environmental neurotoxicants act need to be delineated. Such information will improve prediction and quantification of risks, including risks that become evident only long after exposure.

Neurotoxicologic risk assessment has been largely limited to the application of no-observed-effect levels and uncertainty factors, so it has not generated information on specific risks for given magnitudes of exposure. Experiments should include a range of doses that spans those relevant to expected human exposures. In addition to providing a firmer basis for estimating human risk, such designs permit tests of the assumption that results obtained at high doses predict the pattern of effects at low doses. Such variables as age, sex, duration of exposure, and route of exposure need to be more systematically evaluated. Species-specific effects need to be recognized and investigated. The usefulness of that kind of information for quantitative risk assessment would be greatly amplified by serial measures of both neurotoxic end points and biologic markers. A single model will not be adequate for all conditions of exposure, for all end points, or for all agents. It might be necessary to build risk-assessment models to deal simultaneously with several end points produced by a toxicant. Such models should incorporate biologic markers of neurologic dysfunction and be based on fundamental information on mechanisms derived from experimental test systems and epidemiologic data.

The committee recommends the development of risk-assessment methods that capture the complexities of the neurologic response, including dose-time-response relationships, multiple outcomes, and integrated organ systems.

Experimental designs for studies of neurotoxic agents should provide information needed in the risk-assessment process. Full exploration of relationships and development of risk assessment models would be facilitated by original researchers' making complete data sets available to other investigators.

For the future, one of the major challenges for neurotoxicology will be to use insight from clinical medicine, epidemiology, and toxicology to design effective systems for the prevention of neurotoxic disease in the American public.

Environmental Neurotoxicology

1

Introduction: Defining the Problem of Neurotoxicity

NEUROLOGIC RESPONSES TO ENVIRONMENTAL TOXICANTS

The human nervous system coordinates behavior; in perceiving and responding to external stimuli, it is responsible for mediating communication with the external environment; and it coordinates the activities of all other organ systems and thus plays an essential role in maintaining metabolic balance. The consequences of damage to the nervous system can be profound. Massive injury can result in coma, convulsions, paralysis, dementia, and incoordination. Even slight nervous system damage can impair reasoning ability, cause loss of memory, disturb communication, interfere with motor function, and impair health indirectly by reducing functions, such as attention and alertness, that ensure safety in the performance of daily activities.

Despite the nervous system's compensatory and adaptive mechanisms, many kinds of injury to the nervous system are irreversible, because, after initial development, new nerve cells are not formed; resulting losses of function can be permanent, as well as debilitating. Effects that are difficult to detect in an individual, such as a 5-point decline in intelligence quotient (IQ), are of great concern if

they occur throughout large portions of a population. Prevention of damage to the nervous system is a major objective of social policy, medicine, and public health.

"Neurotoxicity" is the capacity of chemical, biologic, or physical agents to cause adverse functional or structural change in the nervous system. We use the term "environmental neurotoxicity" to refer broadly to adverse neural responses to exposures to all external, extragenetic factors (e.g., occupational exposures, lifestyle factors, and exposures to pharmaceuticals, foods, and radiation); it does not refer merely to the toxic effects of chemicals that are present in the environment as contaminants of air, water, and soil. Table 1-1 lists compounds for which there is evidence of neurotoxicity.

Possible effects of chemical toxicants on the nervous system are varied. For example, Table 1-2 lists neurobehavioral symptoms that—according to clinical reports, epidemiologic investigations, and experimental studies—are caused in humans or animals by at least 25 chemicals (Anger and Johnson, 1985; Anger, 1986). Neurotoxicity can occur at any time in the life cycle, from gestation through senescence, and its manifestations can change with age. The developing nervous system appears to be particularly vul-

9

TABLE 1-1   Partial List of Neurotoxicants

Acetone	Excitatory amino acids
Acetonitrile	Formaldehyde
Acrylamide	Glycerol
Adriamycin	Gold salts
Aliphatic hydrocarbons	Hexane
Alkanes	2,5-Hexanedione
Alkyl styrene polymers	Lead and lead-containing compounds
Aluminum	Isophorone
Ammonia	Lithium grease
N-Amyl acetate	Manganese
Aniline	Mercury and mercury-containing compounds
Antimony sulfide	Methanol
Aromatic hydrocarbons	Methyl acetate
Benzene	Methyl nitrite
Butanol	1-Methyl-4-phenyl-1,2,3,6-tetrahydropyridine (MPTP)
Butyl acetate	1-Nitrophenyl-3-(3-pyridylmethyl) urea
Cadmium	Nitrous oxide
Carbon disulfide	6-OH-Dopamine
Carbon monoxide	Organophosphates
Carbon tetrachloride	Oubain
Chlordane	Ozocerite
Chlordecone	Petroleum distillates
Chlorinated hydrocarbons	Pine oil
Chlorobenzene	Polymethacrylate resin
β-Chloroprene	Products of combustion
Chromium oxides	1-Propanol
Cresol	Propylene gylcol
Cyclohexanol	Pyrethroids
Cyclohexanone	Ricin
Diacetone alcohol	Selenium
o-Dichlorobenzene	Shellac
Dichlorodifluoromethane	Styrene
1,2-Dichloroethane	Tetrachloroethylene
Dichloromethane	Toluene
Dichlorotetrafluoroethane	Trichlorobenzene
Dicylcopentadiene	Trichloroethylene
Dimethylaminopropionitrile	Trichlorofluoromethane
Dinitrobenzene	Tricresyl phosphate
Diphenylamine	Triethyltin; trimethyltin
Dyes	Tungsten oxides
Ergot	Turpentine
Ethanol	Vincristine
Ethyl acetate	Vinyl chloride
Ethylene glycol	Xylene

Source: Adapted from Anger (1986).

TABLE 1-2   Human and Animal Neurobehavioral Effects Attributed to at Least 25 Chemicals

Effect	No. Chemicals or Chemical Groups that Produce the Effect	Effect	No. Chemicals or Chemical Groups that Produce the Effect
Motor		**Affective or personality**	
Activity changes	32	Apathy, languor, lassitude, lethargy, listlessness	30
Ataxia	89	Delirium	26
Convulsions	183	Depression	40
Incoordination, unsteadiness, clumsiness	62	Excitability	58
Paralysis	75	Hallucinations	25
Pupil changes	31	Irritability	39
Reflex abnormalities	54	Nervousness, tension	29
Tremor, twitching	177	Restlessness	31
Weakness	179	Sleep disturbances	119
Sensory		**General**	
Auditory disorders	37	Anorexia	158
Equilibrium changes	135	Autonomic dysfunction	26
		Cholinesterase inhibition	64
Olfaction disorders	37	CNS depression	131
Pain	47	Fatigue	87
Pain disorders	64	Narcosis, stupor	125
Tactile disorders	77	Peripheral neuropathy	67
Vision disorders	121		
Cognitive			
Confusion	34		
Memory problems	33		
Speech impairment	28		

Source: Anger (1986).

nerable to some kinds of damage (Cushner, 1981; Blair et al., 1984; Pearson and Dietrich, 1985; Annau and Eccles, 1986; Hill and Tennyson, 1986; Silbergeld, 1986), but the results of some early injuries may become evident only as the nervous system matures and ages (Rodier et al., 1975).

The observation that some neurologic and psychiatric disorders are of environmental origin is not new (see Table 1-3). Initially, only striking manifestations of neurotoxicity were recognized; e.g., the association of coma, convulsion, and colic with high-dose exposure to lead has been recognized at least since Roman times (Waldron, 1973; Nriagu, 1978). Other long-recognized associations include an increased frequency of depression and suicide among workers in contact with carbon disulfide (Vigliani, 1954); spastic paraparesis (lathyrism) due to

TABLE 1-3 Selected Major Neurotoxicity Events

Year(s)	Location	Substance	Comments
370 B.C.	Greece	Lead	Hippocrates recognizes lead toxicity in mining industry (Klaassen et al., 1986)
1st century A.D.	Rome	Lead	Pliny warns against inhalation of vapors from lead furnaces (Waldron, 1973)
1837	Scotland	Manganese	First description of five cases of chronic manganese poisoning in factory workers handling powdered manganese dioxide (Bellare, 1967)
1924	United States (New Jersey)	Tetraethyllead	In incidents at two plants processing the gasoline additive, over 300 (New Jersey) workers suffer neurologic symptoms and five die; nonetheless, its use in gasoline continues for over 50 years (Rosner and Markowitz, 1985)
1930	United States (Southeast)	TOCP	Compound often added to lubricating oils intentionally added to Ginger (Southeast) Jake, an alcoholic beverage substitute; more than 5,000 paralyzed, 20,000-100,000 affected (Spencer and Schaumburg, 1980)
1930s	Europe	Apiol (with TOCP)	Abortion-inducing drug containing TOCP causes 60 cases of neuropathy (with TOCP)(Spencer and Schaumburg, 1980)
1932	United States (California)	Thallium	Barley laced with thallium sulfate, used as a rodenticide, is stolen and (California) used to make tortillas; 13 family members hospitalized with neurologic symptoms; six die (Spencer and Schaumburg, 1980)
1937	South Africa	TOCP	60 South Africans develop paralysis after using contaminated cooking oil (Spencer and Schaumburg, 1980)
1946	England	Tetraethyllead	People suffer neurologic effects of varied degrees of severity after cleaning gasoline tanks (Cassells and Dodds, 1946)
1950s	Japan (Minamata)	Methymercury	Hundreds ingest fish and shellfish contaminated with mercury from chemical plant; 121 poisoned; 46 die; many infants with serious nervous system damage (Spencer and Schaumburg, 1980)
1950s	France	Organotin	Medication (Stalinon) containing diethyltin diiodide results in more than 100 deaths (Spencer and Schaumburg, 1980)
1950s	Morocco	Manganese	150 ore miners suffer chronic manganese intoxication involving severe neurobehavioral problems (Spencer and Schaumburg, 1980)

Year	Country	Chemical	Description
1956	Turkey	Hexachorobenzene	Hexachlorobenzene, a seed-grain fungicide, leads to poisoning of 3,000-4,000; 10% mortality rate (Weiss and Clarkson, 1986)
1956-1977	Japan	Clioquinol	Drug used to treat travelers' diarrhea found to cause neuropathy; as many as 10,000 affected over 2 decades (Spencer and Schaumburg, 1980)
1959	Morocco	TOCP	Cooking oil deliberately and criminally contaminated with lubricating oil affects some 10,000 people (Spencer and Schaumburg, 1980)
1960	Iraq	Methylmercury	Mercury used as fungcide to treat seed grain is used in bread; more than 1,000 people affected (WHO, 1986)
1964	Japan	Methylmercury	Methylmercury affects 646 people (Spencer and Schaumburg, 1980; WHO, 1986)
1968	Japan	PCBs	Polychlorinated biphenyls are leaked into rice oil; 1,665 people are affected (Goetz, 1985)
1969	Japan	n-Hexane	93 cases of neuropathy follow exposure to n-hexane, used to make vinyl sandals (Spencer and Schaumburg, 1980)
1969	United States (New Mexico)	Methylmercury	Ingestion of pork contaminated with fungicide-treated grain results in severe cases of human alkyl mercury poisoning--first instance of such poisoning in United States (Pierce et al., 1972)
1971	United States	Hexachlorophene	After years of bathing of infants in 3% hexachlorophene, the disinfectant is found to be toxic to nervous system and other systems (Klaassen, 1986)
1971	Iraq	Methymercury	Methylmercury used as fungcide to treat seed grain is used in bread; more than 5,000 severe poisonings and 450 hospital deaths occur; effects on many infants exposed prenatally not documented (Weiss and Clarkson, 1986; WHO, 1986)
1972	France	Hexachlorophene	204 children become ill and 36 die in an epidemic of percutaneous poisoning due to 6.3% hexachlorophene in talc baby powder (Martin-Bouyer et al, 1982)
1973	United States (Ohio)	Methyl n-butylketone (MnBK)	Fabric-production plant employees exposed to solvent; more than 80 workers suffer polyneuropathy, and 180 have less severe effects (Billmaier et al., 1974)

1974-1975	United States (Hopewell, VA)	Chlorodecone (Kepone)	Chemical-plant employees exposed to insecticide; more than 20 suffer severe neurologic problems, and more than 40 have less-severe problems (Spencer and Schaumburg, 1980)
1976	United States (Texas)	Leptophos (Phosvel)	At least nine employees suffer serious neurologic problems after exposure to insecticide during manufacturing process (Spencer and Schaumburg, 1980)
1977	United States (California)	Dichloropropene (Telone II)	24 people are hospitalized after exposure to pesticide Telone II due to traffic accident (CDC, 1978)
1979-1980	United States (Lancaster, TX)	2-t-Butylazo-2-hydroxy-5-methylhexane (BHMH, Lucel-7)	Seven employees of plastic-bathtub manufacturing plant experience serious neurologic problems after exposure to BHMH (Horan et al., 1985)
1980s	United States	1-Methyl-4-pentyl-1,2,3,6-tetrahydropyridine (MPTP)	Impurity in synthesis of illicit drug is found to cause symptoms identical with those of Parkinson's disease (Kopin and Markey, 1988)
1981	Spain	Toxic oil	20,000 persons poisoned by toxic substance in oil; more than 500 die; many suffer severe neuropathy (Altenkirch et al., 1988)
1985	United States and Canada	Aldicarb	More than 1,000 people in California, other western states, and British Columbia experience neuromuscular and cardiac problems after ingestion of melons contaminated with the pesticide (MMWR, 1986)
1987	Canada	Domoic acid	Ingestion of mussels contaminated with domoic acid causes 107 illnesses and three deaths; form of marine vegetation found in estuaries off Prince Edward Island proves to be apparent source of contaminant (Perl et al., 1990)
1988	India	TOCP	Ingestion of adulterated rapeseed oil causes about 600 cases of polyneuritis (Srivastava et al., 1990)
1989	United States	L-trytophan-containing products	Ingestion of a chemical contaminant associated with the manufacture of containing products L-tryptophan by one company results in outbreak of eosinophilia-myalgia syndrome, primarily in the western United States; by 1990, over 1,500 cases have been reported nationwide (Swygert et al., 1990; Belongia et al., 1990)

ingestion of *Lathyrus* species during famines (Denny-Brown, 1947); paralysis following consumption of products contaminated with tri-*o*-cresylphosphate (TOCP), such as the patent medicine Ginger Jake in the United States (Smith et al., 1930) and cooking oil in Morocco (Smith and Spalding, 1959); toxic psychoses in people who inhale tetraethyl lead (Cassells and Dodds, 1946); and erethism (a syndrome with such neurologic features as tremor and such behavioral symptoms as anxiety, irritability, and pathologic shyness) in people exposed to elemental mercury (Bidstrup, 1964).

Numerous other associations between neurologic impairment and environmental exposures have been noted more recently: Exposure to low concentrations of environmental lead is correlated with reduced scores on tests of mental development (Bellinger et al., 1987a,b; Needleman, 1989); the pesticide Kepone (chlordecone) is associated with nervousness, tremors, and other signs of nervous system dysfunction (Cannon et al., 1978); 1-methyl-4-phenyl-1,2,3,6-tetrahydropyridine (MPTP), a byproduct of illicit synthetic-heroin production, causes an irreversible syndrome that closely resembles parkinsonism (Langston et al., 1983); and exposure to manganese can produce parkinsonism and dyskinesia (Cook et al., 1974). In the realm of developmental neurotoxicity, early gestational exposure to ionizing radiation is associated with microcephaly and mental retardation (Otake and Schull, 1984); infants born to women who use cocaine during pregnancy display significant depression of nervous system response to environmental stimuli and other congenital anomalies (Shepard, 1989); heavy ingestion of alcohol by pregnant women produces a syndrome of craniofacial abnormalities and mental retardation in their children (Jones and Smith, 1973), and it is possible that there are persistent adverse effects on mental and motor development in children whose mothers drink more moderately during pregnancy and lactation (Little

et al., 1989; Streissguth et al., 1989; West, 1990).

It has been suggested that exposures to environmental chemicals can contribute to clinical neurodegenerative disorders seen later in life. Calne et al. (1986) have hypothesized that various environmental agents contribute to Alzheimer's disease, Parkinson's disease, or amyotrophic lateral sclerosis (ALS) by depleting neuronal reserves to an extent that becomes observable in the context of natural aging. For example, a syndrome combining the symptoms of ALS and of parkinsonian dementia is prevalent among people from Guam and has been postulated to result from earlier ingestion of one or more chemicals found in the seed of the false sago palm, or cycad (*Cycas circinalis* L.) (Kurland, 1963; Spencer et al., 1987). Some epidemiologic studies have found evidence that various environmental factors are involved in the etiology of ALS, Parkinson's disease, and Alzheimer's syndrome (e.g., Barbeau et al., 1987; Freed and Kandel, 1988; Lilienfeld et al., 1989; Yokel et al., 1988), but other investigations have not detected such relationships (e.g., French et al., 1985; Shalat et al., 1988). The extent to which environmental neurotoxicants contribute to chronic neurologic and psychiatric disease is not yet known.

The nervous system is composed of cells of several types, each with its own functions and characteristic vulnerabilities. Several unique features of the nervous system influence its reactions to toxic agents, including its poor regenerative capacity, its unusual anatomy (especially, long axons), its multiple functions, the extensive interconnections among its cells, its dependence on glucose as an energy source, the existence of highly specialized cellular subsystems, and the wide variety of highly localized neurotransmitter and neuromodulation systems. In addition, environmental toxicants can disturb the complex interactions between the nervous system and other organs. (Chapter 2 discusses the structure and function of the nervous system in some detail, including consideration of the complex neural arrangements involved in information processing and storage.)

The accessibility of a particular part of the nervous system to a specific chemical is a function of both the tissue and the chemical. Many chemicals are kept from entering the brain by the "blood-brain barrier"—the tight junctions formed by endothelial cells surrounding capillaries that supply brain tissue and by endothelial cell-astrocyte interactions. That barrier can, however, be crossed by lipophilic substances and has a series of specific transport mechanisms through which required nutrients, hormones, amino acids, peptides, proteins, fatty acids, etc., reach the brain (Pardridge, 1988). Thus, toxicants can gain entry to the brain if they are lipid-soluble or if they structurally resemble substances that are normally taken up by the nervous system. Moreover, the blood-brain barrier might be less effective in immature than in mature organisms (Klaassen, 1986). It is absent in some parts of the brain, such as in the circumventricular area, and around the olfactory nerve, which runs directly from the nose to the frontal cortex (Broadwell, 1989).

Some cells in the nervous system cannot reproduce themselves; once damaged, they cannot be replaced. Many of those cells ordinarily are present in excess, so there is a buffer against damage, and substantial loss need not affect function or behavior. However, the degree of redundancy, particularly in specific regions, is not known and might vary with age. The presence of such excess can result in a threshold or nonlinear dose-response relationship. Also, the interaction between exposures to neurotoxicants and age-related cell loss might explain delays in the manifestations of toxic consequences. If several processes (e.g., chemical damage, normal aging, and death of cells) are proceeding simultaneously, it might be impossible to isolate a single cause of functional

impairment. A long latent period before toxicity is manifested makes it difficult to associate an exposure and a response causally and even harder to derive a quantitative dose-response relationship.

Manifestations of neurotoxic response can be progressive, with small functional deficits becoming more serious. In those cases, it is almost impossible to define the onset of impairment. For many biologic variables that respond to exogenous agents, the demarcation between an unimportant change and a health-damaging change is unclear. A small change might be a marker of exposure, a moderate change might signal preclinical disease, and a large change might indicate advanced disease.

Some materials, particularly pharmaceuticals, produce different responses in the nervous system at different doses or have adverse side effects at therapeutic doses. For example, the tricyclic antidepressants have a desired therapeutic activity at a low dose, but produce life-threatening anticholinergic effects at higher doses; the antineoplastic drug cis-platinum is a valuable chemotherapeutic agent, but can cause toxic neuropathies; antipsychotic drugs can produce disabling movement disorders; and some antibacterial agents can trigger loss of hearing and balance (Sterman and Schaumburg, 1980). Some substances valued for their relatively selective neuroactivity, such as ethanol, are particularly likely to have simultaneous neurotoxicity (Goldstein and Kalant, 1990).

Exposure to combinations of chemicals can produce interactive effects. Examples include exacerbated hearing loss in persons exposed to some antibiotics (Bhattacharyya and Dayal, 1984; Lim, 1986; Boettcher et al., 1987) and cumulative toxic effects of occupational and environmental exposures to mixtures of solvents (Cranmer and Goldberg, 1986). The general population is exposed to chemicals with neurotoxic properties in foods, cosmetics, household products, air, water, and drugs used therapeutically or recreationally. Exposure to naturally occurring neurotoxins, such as some fish and plant toxins, is yet another aspect of the problem. The multitude of voluntary and unintentional exposures to neuroactive substances that characterize the daily life of ordinary people complicates the task of identifying undesirable neurologic outcomes and attributing them to their proper causes.

MAGNITUDE OF THE PROBLEM OF NEUROTOXICITY

The number of people with neurotoxic disorders and the extent of neurologic disease and dysfunction that result from exposure to toxic chemicals in the environment are not known. However, it is known that an enormous number of people are exposed to environmental materials that in sufficient doses could pose a neurotoxic hazard. The National Institute for Occupational Safety and Health (NIOSH, 1977) estimated that in 1972-1974 there were 197 chemicals to which a million or more American workers were exposed in the occupational setting for all or part of each working day. Anger (1986), in a review of secondary sources in the research literature, found that more than one-third of those 197 chemicals had demonstrated the potential, if the doses were large enough, to produce adverse effects on the nervous system.

The data needed to estimate the overall magnitude of the problem of environmental neurotoxicity do not exist. A National Research Council committee found a few years ago that few of the 60,000-70,000 chemicals in commercial use had been tested for neurotoxicity (NRC, 1984); no information was available on any aspect of the toxicity of approximately 80% of the chemical substances in commercial use. Even among the most extensively regulated classes of chemicals (pesticides, drugs, and food additives), the information needed for a thorough health-hazard assessment was available on only about 5-18%. Few compounds have

been assessed for selective toxicity to vulnerable groups within the population, such as the very young and very old. Moreover, there is little information on the nature and extent of human exposure to even the materials that have been tested and identified as neurotoxic. Finally, not all neurotoxic effects are equivalently damaging or debilitating; scales of relative injury are needed, if the effects of exposures to different materials are to be evaluated in comparable terms.

Given the overall low proportion of substances tested for any type of toxicity and the even smaller fraction that have been assessed for possible neurotoxicity, it is difficult to estimate how many commercial and industrial chemicals are neurotoxic. However, there is evidence that many toxic chemicals in wide use have neurotoxic potential. Of 91 criteria documents produced by NIOSH, 36 (40%) cite neurotoxicity as a reason for recommending limits on occupational exposure (Anger, 1989). Likewise, of 588 chemicals listed by the American Conference of Governmental Industrial Hygienists (A-CGIH) in 1982 as both widely used in industry and having toxicologic significance, 167 (28%) had neurologic effects as one basis for recommendations on maximal exposure concentrations (Anger, 1984). However, O'Donoghue (1986) calculated, on the basis of tests of a small, unselected sample of chemicals, that 5% of *all* industrial chemicals, excluding pesticides, are likely to be neurotoxic. Diener (1987) noted that O'Donoghue's calculation could be an underestimate and should not be considered a firm basis for extrapolation of risk until broader and more systematic surveys of the neurotoxicity of chemicals in commercial use have been undertaken.

In addition to unwanted contact with pollutants or contaminants and inadvertent exposures to industrial products, voluntary exposures to many legal and illegal materials can cause neurotoxic effects (OTA, 1984). Overuse of the legal stimulant caffeine can lead to tremors, irritability, and sleeplessness. Exposure to the depressant ethanol is responsible for many accidents on the high-way and elsewhere, fetal injuries, chronic disease, and antisocial behavior. The devastating personal and societal consequences of "substance abuse," engaged in at least initially for psychoactive effects, have become obvious. The list of self-administered neuroactive substances is extensive, and exposure to them makes epidemiologic identification of additional neurotoxicants difficult.

The Health Care Financing Administration of the U.S. Department of Health and Human Services has estimated that $23 billion was spent in 1980 for the care of people with diagnosed neurologic diseases; in many cases, these illnesses are due to accidental exposures to neurotoxicants or to use of legal drugs, rather than to use of illegal substances (Rice et al., 1985). As has been suggested regarding lead toxicity (Provenzano, 1980; EPA, 1985), it would not be surprising if the direct and indirect societal costs of subclinical neurologic losses or deficits—such as reduction in intelligence, diminution in achievement, and waste of opportunity—might equal or exceed those of neurotoxic effects that are clinically recognized. Modifications of behavior induced by neurotoxic substances can also adversely affect people other than those directly experiencing the toxic effects, as is the case for victims of drunken drivers.

DETECTION AND CONTROL OF EXPOSURE TO NEUROTOXICANTS

Prevention is the key to dealing with neurologic diseases of toxic environmental origin. Such diseases result from exposures to synthetic and naturally occurring chemical toxicants encountered in the ambient environment, ingested in foods, or administered as pharmaceutical agents. Many of the diseases are not curable, so they must be prevented. They can be prevented by eliminating or reducing exposures at the source (primary prevention), or they can be controlled by early detection and diagnosis of neurotox-

ic effects while they are still reversible or while they are at a relatively early stage of evolution when their progression can still be halted (secondary prevention).

In light of the difficulty of regulating potentially neurotoxic compounds found in illicit substances of abuse or in certain foods, particularly in plants, the most effective approach to preventing neurotoxic disease of environmental origin consists of identification of neurotoxicity through routine pre-market testing of all new chemicals before they are released and before human exposure has occurred. Disease is prevented by restricting or banning the use of chemicals found to be neurotoxic or by instituting engineering controls and imposing protective devices at points of environmental release and potential human exposure.

Given the current absence of data on neurotoxicity of most chemicals in commerce, an extensive program of primary prevention through toxicologic evaluation is needed. The Environmental Protection Agency (EPA) has regulatory authority to screen chemicals coming into commerce, but some have expressed the opinion that the authority is insufficiently exercised and inadequately supported. Therefore, people might be overexposed to environmental neurotoxicants and might develop neurotoxic responses to chemicals that slip through the regulatory net. For example, important side effects of pharmaceutical agents continue to surface, such as the 3 million cases of tardive dyskinesia in patients on chronic regimens of antipsychotic drugs. Prudence dictates that all chemical substances—both those already on the market and new ones—be considered potential neurotoxicants; a chemical cannot be regarded as free of neurotoxicity in the absence of data on its toxicity. Prudent public policy therefore dictates that all chemicals, both old and new, be subjected to at least basic screening for neurotoxicity when use and exposure warrant.

Neurotoxicants can be identified by several means, ranging from clinical case reports and observations by alert physicians to formal in vivo and in vitro screening. Analysis of structure-activity relationships (SARs) has been used extensively to predict neurotoxicity, but SARs as currently used can provide only minimal guidance for hazard identification. EPA, however, uses SARs as a basis for decisions on testing under the Toxic Substances Control Act. Testing procedures designed for in vivo neurotoxicologic evaluation of new chemicals have been developed and are reasonably effective, but they are tedious and labor-intensive. Additional mechanistically based in vitro test systems are badly needed, and they must be correlated with in vivo test systems. In summary, a rational and efficient strategy for neurotoxicity testing is needed to close the gap in neurotoxicity testing. Development of a blueprint for such a strategy was a major objective of the study reported here.

SCOPE OF THIS REPORT

The Committee on Neurotoxicology and Models for Assessing Risk was formed at the request of the Agency for Toxic Substances and Disease Registry to review the biologic principles and mechanisms of neurotoxic action for the purpose of using them in risk assessment. This report deals with environmental neurotoxicology—with the current state of the field, its implications for human health, and the prospects for its rational future development. The committee first reviewed current knowledge of environmental neurotoxicity, paying special attention to the types of neurologic injury caused by chemicals in the environment, to determine the overall extent of the problem of environmental neurotoxicity. We investigated ways in which improved biologic markers of neurotoxicity (markers of subtle and subclinical effects, of exposure to neurotoxicants, and of

susceptibility to their effects) could be developed, could improve laboratory tests, and could facilitate the early recognition of neurotoxic injury in exposed human populations. We then considered available methods of testing for neurotoxicity and evaluated strategies for recognizing and characterizing the neurotoxic potential of chemicals and for detecting neurotoxic effects among environmentally or occupationally exposed humans. Finally, we examined approaches to neurotoxicologic risk assessment.

This chapter has provided an overview of the problem of environmental neurotoxicity, described the level of existing knowledge, and suggested broad issues for scientific research. Chapter 2 illustrates the biologic basis of neurotoxicity by reviewing the aspects of the structure and function of the nervous system that are subject to disruption by exogenous agents. Chapter 3 discusses biologic markers, which provide a conceptual thread connecting human, animal, and in vitro studies. Chapter 4 considers approaches to identifying neurotoxicants and describes available in vitro and in vivo tests that could be used to assess toxicity before human exposure has occurred. It develops a blueprint for a multitiered program to test chemical substances for neurotoxicity, and it lays out a plan for systematic evaluation of the results of neurotoxicity testing. Chapter 5 describes approaches to the epidemiologic surveillance of human populations at risk of environmental neurotoxicity and outlines the uses to which human data can be put in bringing neurotoxic exposures under control. And Chapter 6 describes current approaches to risk assessment and discusses modifications in the risk-assessment paradigm that will be needed to accommodate neurotoxic end points. And Chapter 7 presents the committee's conclusions and recommendations.

2

Biologic Basis of Neurotoxicity

The nervous system is the communication and information-storage system of the body. In addition to coordinating functions that are commonly associated with the brain, such as thinking, the nervous system plays a major role in physiologic regulation. Virtually all physiologic functions are influenced or controlled by the nervous system. Other systems, such as the circulatory and reproductive systems, provide information to the nervous system, which in turn controls their functioning. The nervous system also integrates the functions of the various organ systems. Thus, neurologic dysfunction has a major impact that goes beyond the nervous system itself. Conversely, abnormal functioning in other systems can modify neurologic functioning.

Our abilities to act, to feel, to learn, and to remember all depend on the complex anatomic arrangements and proper integrative functioning of the nervous system, of which we have only a rudimentary understanding. The ultimate goal of neuroscience is to understand how this system functions normally and how subtle perturbations can lead to disease and dysfunction. Observation of the response of experimentally exposed animals or inadvertently exposed humans to neurotoxic chemicals advances our under-

standing of neurobiology while generating information about the consequences of exposure to environmental pollutants.

The first two sections of this chapter summarize our current understanding of the nervous system. The next section discusses several aspects of the nervous system that make it especially vulnerable to chemically induced damage. The fourth section describes examples of neurotoxicant-induced damage of several types. A final section summarizes the problems arising from the nature of neurobiology that must be addressed when a substance is tested for neurotoxicity, the subject of Chapter 4.

CELLULAR ANATOMY AND PHYSIOLOGY

The nervous system is composed of two general types of cells, nerve cells (neurons) and glial cells, which are present in approximately equal numbers. Neurons are responsible for the reception, integration, transmission, and storage of information. They can be classified as interneurons and motor, sensory, and neuroendocrine neurons. Motor neurons innervate and elicit contraction of muscle fibers. Sensory neurons monitor the organism's external and internal environments. Specialized receptors on sensory nerves convey information regarding light, sound, muscle position, taste, and smell to the central nervous system. In many cases, sensory neurons form the input (afferent) side of information loops in the central nervous system; the output (efferent) side is mediated by motor fibers or by the endocrine system. Neuroendocrine neurons elicit secretions from glandular cells. Interneurons facilitate or modulate interactions between other neurons, often in complex feedback loops. Glial cells differ from neurons in their ability to divide throughout the organism's life, particularly in response to injury, and in their functions (Table 2-1), which are less well understood than those of neurons, but are believed to be largely restricted to structuring and regulating the environment surrounding the neurons. Glial cells also play a major role in myelination.

Structure of the Neuron

A neuron (Figure 2-1) has a more or less round cell body (soma), which contains its nucleus and organelles for protein synthesis. The diameter of the cell body of a vertebrate animal is about 10-100 μm (the largest would be roughly the size of the period at the end of this sentence). The neuron gets its distinctive appearance from its dendrites and axon, elongated processes that emanate from the cell body. The length and branching patterns of the dendrites vary from one type of neuron to another. Dendrites and their branches greatly increase a neuron's surface area, and this increases the number of messages that the neuron can receive and the number of sources from which they can come. The arrangement is analogous to the branching of a tree, which increases its reception of sunlight by increasing its total leaf surface area.

In most neurons, the axon is a thin, particularly elongated cellular process emanating from the neuronal soma and is specialized for the conduction of nerve signals away from the cell body and toward its synapse with other cells (other neurons, muscle cells, or gland cells). One neuron communicates with another by releasing chemical signals onto a small gap, or the synapse, between the tip of an axon and the dendrites or soma of its target cell. Axons are shorter or longer, depending on the type of neuron, but an axon's length is generally tens or thousands of times greater than the diameter of the cell body. An extreme example is the axon of a motor neuron that innervates the

TABLE 2-1   Nonneuronal (Glial) Cells of the Nervous System and their Function

Glial Cell Type	Functions
Astrocyte	Repair, structural support, transport
Oligodendrocyte	
Satellite	Neuronal maintenance
Intermediate	Potential for maintenance or myelination
Intrafasicular	Myelination
Microglia	Phagocytosis
Ependymal cells	Service buffer between brain and cerebrospinal fluid
Schwann cells	Myelination of peripheral nerves

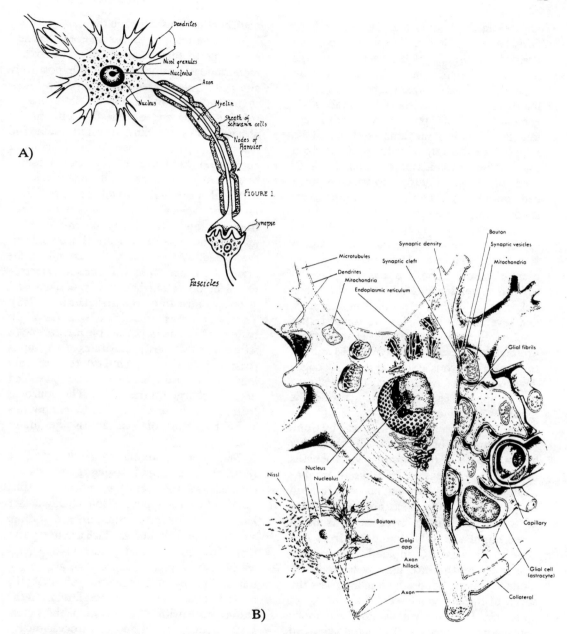

FIGURE 2-1   Diagrammatic representation of neuronal structure. A) The drawing illustrates the gross structure of a neuron with its many dendrites, myelin sheath, and synaptic contact with another neuron. Source: Liebman, (1991). B) The drawing shows structures visible at the electron microscope level. The nucleus, nucleolus, chromatin, and nuclear pores are represented. In the cytoplasm are mitochondria, rough endoplasmic reticulum, the Golgi apparatus, neurofilaments, and microtubles. Along the surface membrane are such associated structures as synaptic endings, astrocytic processes, and a capillary containing a red blood corpuscle. Source: Willis and Grossman, (1973).

muscles in the human toe. Its cell body (a fairly large one) lies in the spinal cord at the level of the middle of the back; its axon is more than a meter long. Near its target, an axon divides into many small branches called telodendria. Telodendria end in small enlargements called synaptic boutons, which contain the chemical messengers that will be released to the target cell. A single axon can end in thousands or hundreds of thousands of synaptic boutons, by which a nerve can communicate with a few or many thousands of target cells.

Many axons are insulated by myelin, the extended and modified plasma membrane of specialized glial cells. During early development, myelin is deposited in a spiral fashion to form a sheath around the axon. Myelin is richer in lipid than are other membranes, and several unusual lipids are found in the myelin sheath; some well-characterized proteins are peculiar to myelin. Myelin of the central nervous system is made by oligodendrocytes and differs from the myelin of the peripheral nervous system, which is made by Schwann cells.

The myelin sheath formed by one oligodendrocyte or Schwann cell covers only a short length of the axon; adjacent lengths of axon are ensheathed by adjacent myelinating cells. Very short lengths of bare axon between myelinated areas are called nodes of Ranvier. Within myelinated axons, a nerve impulse travels by jumping from one node to the next. Compared with conduction in axons that are not myelinated, this mode of travel, called saltation, increases the speed of impulse conduction by a factor of up to 100. In a poorly understood metabolic dependence between a myelinated axon and its ensheathing cells, some types of injury to one cause dysfunction in the other.

The Nerve Impulse

Two processes are involved in communication by neurons: electric (discussed here) and chemical (discussed in the next section). Within a neuron, transmission is electric and analogous to that in a battery. The axonal membrane is semipermeable to the positively and negatively charged ions (chiefly, potassium, sodium, and chloride) that are in solution inside and outside the cell. In the absence of active and directed metabolic transport, those ions would be distributed equally on each side of the axonal membrane, and the inner and outer electrochemical potentials would be the same. However, several enzyme systems normally transport ions actively from the inside of the axon to the outside or from the outside to the inside. As a result, the axonal membrane separates areas of unequal electrochemical potential; i.e., it is polarized and generates a resting potential (Davies, 1968). The situation is analogous to a flashlight battery, where an impermeable membrane (a strip of cardboard) separates two fluids (pastes in a dry-cell battery) that contain ions of positive charge on one side and negative charge on the other. The neuronal "battery" must be maintained by continuous active transport of ions across the membrane.

The action potential is generated as a result of opening and closing of various ion channels in the nerve membrane (Hille, 1984). Several types of ion channels are known, but the most common are sodium and potassium channels, which are tiny pores that allow sodium and potassium ions, respectively, to pass through. The action potential is propagated regeneratively along the nerve fiber without loss of amplitude. In the resting state without stimulation, the potassium channels open and close occasionally, and the sodium channels remain mostly closed. Therefore, the resting nerve membrane is permeable mainly to potassium, but only sparingly to sodium. That results in the resting potential, with the cell's interior being negative with respect to the outside by 70 mV or so. On electric stimulation, however, many sodium channels start opening, and that makes the membrane permeable predominantly to sodium. Electric stimulation thus generates an action potential with a peak some 50 mV more positive on the interior than the exterior of the nerve. Soon the sodium channels start closing and the potassium channels start opening at a high frequency; this brings the membrane potential back to the resting level. During the action potential, sodium ions enter the cell through open sodium channels and potassium ions leave the cell through open potassium channels; the result is a slight imbalance of the internal sodium and potassium concentrations, which is corrected quickly by a metabolic pump that extrudes extra sodium and absorbs potassium. Thus, the cell is capable of generating many action potentials in rapid succession.

The ion channels are the primary target sites of some toxicants and therapeutic drugs. For example, tetrodotoxin (a puffer fish poison) and local anesthetics block the sodium channel, thereby impairing the impulse conduction. The insecticidal pyrethroids and DDT keep the sodium channels open for unusually long periods, making the nerve generate impulses in a hyperactive fashion.

Synaptic Transmission

When a nerve impulse reaches a terminal branch of an axon, it depolarizes the synaptic boutons. That initiates the second type of transmission by means of the cellular release of special messenger chemicals (neurotransmitters) that are stored in the boutons. The amount of secretion is related to the degree of depolarization, which, in turn, is a function of the number of nerve impulses that reach a synaptic bouton. A very narrow gap, the synapse, separates the bouton from its target cell (Figure 2-2). The released chemicals diffuse across the synapse and attach themselves to specific receptor molecules embedded in the membrane of the target cell. Attachment (binding) initiates one of several responses, usually involving changes in ion flux, depending on the type of target cell and the type of messenger-receptor pair involved. For example, when the target cell is a muscle cell, the neurotransmitter-receptor interaction leads to contraction of the muscle; when the target is a gland cell, the interaction leads to secretion.

Synaptic transmission between neurons is slightly more complicated. Messenger molecules from the bouton of a presynaptic (releasing) cell bind to receptors on the surface of the postsynaptic (receiving) cell; that can lead to either excitation or inhibition of the target neuron. Neuronal cell bodies and dendrites are polarized in the same way as the axon. At an excitatory synapse, the neurotransmitter-receptor interaction leads to a momentary opening in ion-specific channels that produces depolarization (excitatory postsynaptic potential). At an inhibitory synapse, a different neurotransmitter-receptor interaction leads to a momentary opening of different ion-specific channels that produces an increase in polarization (hyperpolarization) or a small depolarization (inhibitory postsynaptic potential). Unlike the membrane of the axon, the membrane of the cell body and dendrites does not generate nerve impulses. Instead, the sum of all the excitatory and inhibitory postsynaptic potentials determines a moment-by-moment average polarization that is monitored by the portion of the cell's axon adjacent to the cell body. When the running average exceeds a threshold depolarization at the axon's initial segment, a nerve impulse is generated and begins its electrochemical journey down the axon.

The duration of neurotransmitter action is a function primarily of the time that it remains in the synapse. Under normal circumstances, that is a very short time (a few thousandths of a second), because specialized enzymes quickly degrade the neurotransmitter, and, in some cases, metabolically driven re-uptake systems rapidly scavenge

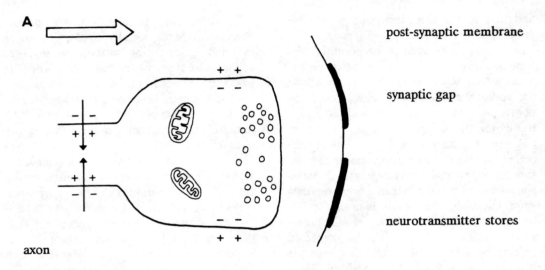

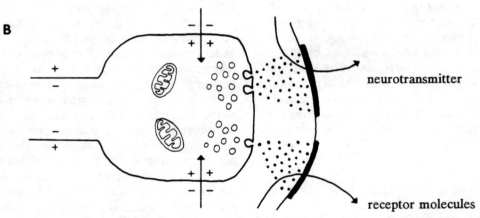

FIGURE 2-2   Events in chemical synaptic transmission. **A**, action potential shown approaching axon
terminal containing transmitter stored in synaptic boutons. **B**, arrival of nerve impulse triggers
release of transmitter into synaptic gap. Transmitter diffuses across synapse and reacts with
receptor molecules associated with postsynaptic membrane. Resulting change in membrane
permeability to ions, such as sodium and potassium, causes flow of synaptic current. In this case,
current flows inward through postsynaptic membrane and results in depolarization called excitatory
postsynaptic potential. Later transmitter action is terminated by enzymatic destruction, diffusion,
or re-uptake. Source: Willis and Grossman (1973).

the synapse for neurotransmitter and transport it back into the synaptic bouton of the
releasing cell.

Synaptic Messengers

Several   dozen   neurotransmitters   are

known, and more are discovered every year.
(A partial list is given in Table 2-2.) A given
neuron can synthesize and release a single
neurotransmitter or, in the case of peptidergic neurons, several neurotransmitters. Similarly, a neuron can contain receptors for and
be influenced by several neurotransmitters.

TABLE 2-2 Neuron Type Classified by Neurochemical Released for Synaptic Transmission

Neuron Type	Neurotransmitter
Cholinergic	Acetylcholine
Excitatory amino acid	Glutamate, aspartate
Inhibitory amino acid	γ-Aminobutyric acid (GABA), glycine
Catecholaminergic	Dopamine, epinephrine, norepinephrine
Sertoninergic	Sertonin
Peptidergic	Substance P, enkephalins, endorphins, neurotensin, carnosine, angiotensin II, pituitary hormones and their hypothalmic controlling factors
Other	Histamine, taurine, proline, N-acetyl-L-aspartate, serine, octop amine, p-tyramine, N-acetyl serotonin, tryptamine, purine, 5-methoxytryptamine, pyrimidine derivatives

Research on the functions of the multitude of synaptic messengers is one of the most active fields in neuroscience today. Although not all the details are known, one conclusion is clear: neurons communicate by a complex code that involves a large number of chemicals and exquisitely specific chemical interactions.

GENERAL ASPECTS OF NERVOUS SYSTEM STRUCTURE AND FUNCTION

The nervous system has two parts: the central nervous system (CNS) and the peripheral nervous system (PNS). The CNS comprises the brain and the spinal cord. The PNS is composed of nerve cells and their processes that lie outside the skull and vertebral column. The nerve cells of the PNS are generally found in aggregates called ganglia. The neurons of the CNS are often segregated into aggregates called nuclei. Various nuclei are functionally related to one another to constitute levels of organization called systems; thus, we speak of

the motor system, the visual system, and the limbic system.

The cells of the nervous system are organized in networks of increasing evolutionary complexity. In anatomic terms, the most primitive level is the spinal cord, wherein responses to sensory signals are virtually instantaneous; such responses are called reflexes. Spinal-cord functions tend to have been conserved in evolution, so the system in lower vertebrates resembles that in humans. The oldest regions of the brain regulate involuntary activities of the body, such as breathing and blood circulation, and exert much of their influence via the reflexes of the spinal-cord. The brain stem—which comprises the medulla, pons, and midbrain—has functions related primarily to motor and sensory activities of the head and neck, including hearing, balance, and taste. The diencephalon is composed of the thalamus and the hypothalamus. The thalamus processes most information from the rest of the nervous system on its way to the cerebral cortex, and the hypothalamus regulates autonomic, endocrine, and visceral integration. The basal ganglia assist the cerebral cortex with motor functions. The cerebral cortex, the most highly evolved level, is a vast information-storage and -retrieval area.

In addition to its role in such functions as perception and learning, the nervous system has a critical role in controlling the body's endocrine glands. That control is concentrated in several small nuclei at the base of the brain that are known collectively as the hypothalamus. Nerve cells in the hypothalamus secrete chemical messengers into a short loop of blood vessels that carries them to the pituitary, the body's master gland. The pituitary, in turn, releases chemical messengers into the general circulation. The pituitary messengers regulate other glands, such as the thyroid, the adrenals, and the gonads. The entire system maintains homeostasis, i.e., a regulated state of optimal physiologic functioning or regular cycling throughout the body.

Metabolism

The CNS has a very high rate of metabolism. In an average adult human, the weight of the brain (about 3 lb) is approximately 2% of total body weight. But its energy demands are so great that it receives about 14% of the heart's output and consumes about 18% of the oxygen absorbed by the lungs (Milnor, 1968). Unlike other organs, the brain depends almost entirely on glucose as a source of energy and as a raw material for the synthesis of other molecules (Damstra and Bondy, 1980). The CNS depends critically on an uninterrupted supply of oxygen and on the proper functioning of enzymes that metabolize glucose. Other tissues (muscle, for example) have alternative sources of energy and can survive for a relatively long time when the circulation is compromised, but nerve cells usually begin to die within minutes.

Blood-Brain Barrier

With few exceptions (e.g., proteins), substances that are carried in the bloodstream can pass from capillaries to surrounding extracellular fluid (in nearly all regions of the body). That is not true in the central nervous system (Betz et al., 1989). The cells that form the walls of capillaries in the CNS abut one another with unusually tight junctions. Those tight junctions, and perhaps other cells that line the outside of the capillaries, form a barrier that prevents the free passage of most bloodborne substances and thereby helps to create fine control of the extracellular environment of nerve cells. (Peripheral nerves and the retina have a similar barrier.) Without this blood-brain barrier, the brain would be even more vulnerable to chemical damage.

Chemical access to cells varies in different regions of the nervous system and at different stages of development (Klaassen, 1986). The area postrema and circumventricular areas have no blood-brain barrier. In developing animals, monosodium glutamate selectively kills cells in the arcuate nucleus and the retina, where the barrier is insufficient to exclude it. Other areas have incomplete barriers; e.g., the dorsal root ganglion is vulnerable to penetration and resulting cell-killing by chemicals (such as doxorubicin) that are excluded from the nearby spinal cord by a more complete barrier (Cho et al., 1980), and the olfactory nerve directly connects the outside environment via the nose to the frontal cortex (Broadwell, 1989).

Most nuclei in the nervous system, particularly in the CNS, have relatively intact barriers and thus are protected from some classes of chemicals. Although those barriers exclude many chemicals from the brain, some neurotoxic chemicals can gain access to the CNS via normal uptake mechanisms, if they are structurally similar to endogenous substances normally taken into the brain.

Such variable protection places some anatomic regions at particular risk and gives rise to different expressions of neurotoxicity in response to various agents.

As will be discussed in more detail in Chapter 3, the selectivity of the blood-brain barrier results in fairly specific biologic markers that correspond to particular substances. Because the CNS is enclosed in a fairly isolated compartment defined by the blood-brain barrier, monitoring of more readily sampled tissues, such as blood, is not usually sufficient for detecting biologic markers of exposure or effect in the CNS itself; for that purpose, it might be necessary to sample spinal fluid.

Development

The organizational complexity of the adult human nervous system is staggering. Indeed, the complexity is so great that we have only a rough idea of the number of neurons in the brain. The most widely cited estimate is 30 billion. It is reasonable to assume that an average neuron receives input from about 1,000 other neurons and, in turn, sends its messages to another 1,000 neurons.

Details of the development of the human nervous system are poorly understood, but the broad outline is known. It begins early in embryonic development and is not complete until about the time of puberty. It proceeds as a carefully timed multistage process guided by hormones and other chemical messengers. Some of the initial organization acts as scaffolding, i.e, a temporary framework for later development. The temporal and spatial organization of the developmental process is precise and complex, and the deleterious effects of a relatively minor disturbance—one that perturbs developmental interactions among just a few cells and for only a brief time—might be amplified enormously through the cascade of developmental steps. Developmental stage

is important in determining the limits of plasticity, i.e., the relocation of cellular resources to overcome the effects of injury; this ability to adapt to damage is thought to be greatest early in development. For example, a lesion that might be compensated for in a developing nervous system might produce irreversible deficits in an adult. In contrast, interference with a particular cell-cell interaction can have qualitatively different effects in developing and mature nervous systems. For example, dopamine receptors respond to a decrease in dopamine input during development with a permanent decrease in their number, but in adulthood with an increase in their number (Rosengarten and Friedhoff, 1979; Miller and Friedhoff, 1986). Thus, the nervous system's developmental period is a time of both vulnerability and ability to compensate.

Disuse of a particular synaptic connection can result in its inactivation and eventual dissolution. When more than a critical number of synaptic connections on a particular neuron are inactivated, the neuron itself might become superfluous and be deleted. That phenomenon occurs extensively during development: excess neurons and synaptic connections are established, many of which apparently disappear during the modeling of the mature nervous system. The converse also happens: synaptic connections used repeatedly become fixed, and transmission at these locations is facilitated through long-lasting changes in how ion channels or receptors are linked to second messenger systems. It appears that the shape of a given synapse is plastic and that the arrangement of structural proteins within a synapse can be modified to alter the efficiency of information transmission.

Neuron production does not cease at birth in the human (or in any common laboratory species), but continues through most of the first year of postnatal life. However, neurogenesis for most regions is completed during embryonic stages or fetal stages. Thereafter, neurons that die are not replaced, although supporting cells continue to divide. Later development, maturation, or repair of the nervous system involves extension of the population of neuronal processes (but not production of new neurons) and modification of interconnections. That is in marked contrast with most other tissues, in which cell replacement is continuous. The absence of neurogenesis in the adult places severe constraints on the nervous system's ability to recover from damage.

Plasticity

Although the general pattern of interconnections in the nervous system is fixed according to the genetic program that guides development, the fine structure of the pattern of connections in at least some regions probably responds to exogenous factors and retains a considerable degree of adaptability throughout life. This characteristic of plasticity involves rearrangement of the details of synaptic connections. Plasticity confers on the CNS the capacity to respond to environmental signals by learning and memory formation. Damage to the nervous system can also evoke rearrangement of connections among neurons that survive the insult; rearrangement permits a limited, but important, functional adjustment to compensate for neuronal damage.

The clearest example of neuronal plasticity is seen after damage to axons in peripheral nerves. Neurons supplying peripheral nerves usually survive when their axons have been damaged. Such a neuron is able to regenerate its axon, which then locates its original target. That is the basis, for example, of the eventual partial return of sensation and muscle control in a surgically reattached limb. Neurons in the CNS are also able to regenerate damaged axons, but it is far more difficult for them to reacquire their targets, partly because of the barrier formed by proliferating glia, partly because of the inherently more complex interconnections of the CNS, and partly because the CNS lacks a pathway for regenerating axons to follow, such as is formed in the PNS by basal laminae and collagen and reticular fibers of the endoneurium (the tissue that separates the fibers in a peripheral nerve).

Trophic Interactions

The functioning of individual neurons is determined partly by a continuous supply of trophic factors, chemicals that originate in the cell's targets. For example, if the axons of motor neurons are severed, both the motor neuron and the muscle cells atrophy because of the lack of the chemicals that they normally exchange. A target provides trophic feedback to the neuron that innervates it, so compromise of neuronal function can be secondary to injury to nonneuronal cells. Such interdependence is also seen among neurons. Some neurons atrophy when the neurons that normally communicate with them die (Kelly, 1985). Thus, it is possible that damage can spread from a small number of inactivated neurons.

VULNERABILITY OF THE NERVOUS SYSTEM TO CHEMICAL TOXICANTS

In general terms, neurotoxic injury involves adverse functional or structural change. But the different vulnerabilities of various cells to injury by a given toxicant and the normal roles of susceptible cells determine the biologic markers of effect or clinical manifestations of exposure to a particular neurotoxic chemical. For example, myelin degeneration after exposure to triethyltin or hexachlorophene is manifested as spasticity; effects on neurons in the hippocampus due to exposure to trimethyltin, triethyllead, or methylmercury are associated with learning and memory deficits; degeneration of dopamine neurons in the substantia nigra due to MPTP exposure leads to the tremor and rigidity of parkinsonism; and axonal degeneration in the PNS due to γ-diketones or acrylamide is reflected in muscular weakness and sensory deficits. Such effects can often be detected in either exposed humans or corresponding experimental animal models with clinical examination or other diagnostic means, such as electrophysiology, computed axial tomography and positron-emission tomography (CAT and PET scanning), and magnetic resonance imaging. In contrast, some of the acute effects of toxic chemicals are manifested as dysfunction of the nervous system without any observable structural changes; examples are intoxication by insecticide pyrethroids, DDT, and malathion. Several factors render the nervous system especially vulnerable to chemical assault; they are described briefly in the next few pages.

Complexity of Structural and Functional Integration

The proper functioning of the nervous system depends critically on complex interactions among different cell types in many anatomic locations that communicate via a variety of electric and chemical signals. The sheer complexity of the system makes it vulnerable, just as the functioning of a computer is more vulnerable to damage than is the functioning of an abacus. Other organs also are complex and have many cells, but most have a high degree of functional redundancy. For example, the loss of a few liver cells creates little disturbance in the organ's function. In contrast, although the nervous system is able to compensate extensively after injury, its functional redundancy is nonetheless circumscribed. A small lesion might initially result in a functional deficit (for example, a small stroke might impair speech or motor coordination), but eventually be compensated for as new synaptic pathways are established. If a lesion is widespread, the corresponding functional capacity might be lost; for example, a relatively small loss of neurons that use acetylcholine as their neurotransmitter might produce a disturbance of memory, whereas a relatively minor insult concentrated in a subsystem that relies on dopamine as its neurotransmitter might have no effect, and a larger insult might impair motor coordination.

Limitations on Repair

To understand neurotoxicity, one must remember that adults have no neurogenesis. Nerve cells that are destroyed by chemical insult are not replaced; the lesion is permanent. However, the nervous system has remarkable compensatory mechanisms. Such compensation is evident, for example, in the development of tolerance to morphine and alcohol, the long-term potentiation phenomenon documented in experimental animals, functional recovery after stroke or brain surgery, and neuronal sprouting after axonotomy. However, the scantiness of functional redundancy and the lack of neurogenesis limit (although they do not preclude) compensatory ability, particularly after extensive neuronal death. And the limits are exacerbated by normal aging processes: Neurons die throughout adult life, and additional cell loss due to exposure to toxicants may cause a further deficit in the functional reserve that is ordinarily drawn on in aging.

Accessibility to Lipid-Soluble Toxicants

All cell membranes are composed chiefly of lipid molecules. However, the lipid content of the nervous system is especially high—the myelin in axonal sheaths, for example, constitutes approximately 25% of the brain's dry weight. Many classes of toxicants, such as chlorinated hydrocarbon pesticides and industrial solvents, dissolve readily in lipids. Molecules dissolved in membrane lipids are relatively protected from enzymatic degradation and sheltered from the general circulation. Thus, lipids act as depots where various toxicants can accumulate and reach high local concentrations. Toxicants can gradually leach out of lipid depots, such as liver and adipose tissue, to such an extent that the exposure of the nervous system itself to a toxic chemical lasts much longer than the external exposure of the organism (Boylan et al., 1978). Toxicant concentrations in adipose tissue can reflect exposure of other lipid-rich tissue, such as is found in the nervous system, but determination of the temporal course of exposure or of tissue doses in the CNS is more problematic.

Dependence on Glucose

The nervous system depends almost exclusively on glucose for energy and as a starting material for the synthesis of other molecules (Sokoloff, 1989). Its energy requirements, e.g., for maintaining electric potentials for meeting the bioenergetic demands of neurotransmitter reuptake, and for performing axonal transport are high. The system is therefore particularly vulnerable to chemicals that inhibit the enzymes that metabolize glucose.

Axonal Transport

All types of cells must transport proteins and other molecular components from their site of production near the nucleus to other sites in the cell. Intracellular transport is relatively simple in most cells, because the greatest distance to be traversed is only a small fraction of a millimeter. Neurons are unique among cells, in that the cell body not only must maintain the functions normally associated with its own metabolic support, but also must provide support to dendrites

and axons, often over relatively vast distances. The axon and synaptic boutons have little ability to synthesize the materials needed to sustain their structure and function, so they depend on delivery of the materials from their parent-cell body (Hammerschlag and Brady, 1989). Delivery of such substances by intracellular transport down the axon (axonal transport) is highly vulnerable to interruption by toxic chemicals. In addition, the functional integrity of the neuronal cell body often depends on a reciprocal supply of trophic factors from the cells that it innervates, and these trophic factors must also be supplied via axonal transport over long distances with very high energy requirements.

Axonal transport has several forms. Fast anterograde (forward) transport carries macromolecular assemblies of primarily membrane-associated glycosylated or sulfated proteins along structural tracks of microtubules from cell body to distal axon at a rate of about 400 mm/day. From the cell body, slow anterograde transport carries soluble enzymes involved in metabolism and neurotransmission at 1-2 mm/day in conjunction with microtubules and neurofilaments or at 5-10 mm/day in association with microfilaments. Fast retrograde (backward) transport carries a variety of materials up the axon to the cell body at about 250 mm/day; its cargo might be endogenous molecules (largely glycoproteins) destined for recycling, trophic signals, or exogenous materials (e.g., tetanus toxin, lead, and doxorubicin, which might circumvent the blood-brain barrier by traveling up the axon from the PNS).

It has been hypothesized that the great demand of axonal transport for energy derived from oxidative metabolism is a point of neural vulnerability to some toxicants. Toxicants that interfere with that metabolism or that disrupt the spatial arrangement or manufacture of neurofilaments block axonal transport. Many chemicals that produce distal axonopathies probably act by such mechanisms. For example, γ-diketones (metabolites of *n*-hexane and methyl *n*-butyl ketone) appear to exert their toxic effect via the mechanism of neurofilament cross-linking (St. Clair et al., 1989), whereas several biochemical mechanisms (e.g., reaction of sulfhydryl moieties and altered glycolysis) might contribute to acrylamide's impairment of axonal transport (Miller and Spencer, 1985).

Synaptic Transmission

Neurons communicate with one another and with muscle and gland cells via a large number of chemical messengers that interact in precisely defined ways with specialized receptor molecules. Chemically mediated communication is vulnerable to disruption by exogenous chemicals.

Disruption can occur in several ways. For example, it is functionally important for synaptically released neurotransmitters to have brief effects. Some neurotoxicants, such as the organophosphate pesticides, inhibit the enzyme that terminates the effect of the neurotransmitter acetylcholine on its target. The result is an injurious overstimulation of the target cell. Other neurotoxicants, particularly toxins and their chemical relatives, mimic the action of a neurotransmitter by interacting with its receptor molecule. For example, LSD (lysergic acid diethylamide), a synthetic relative of a natural toxin (ergot) produced by a fungus that infects cereal grain, interacts with the receptor for the neurotransmitter serotonin (Aghajanian, 1972).

Other neurotoxic chemicals act by deranging other aspects of chemical neurotransmission. Some chemicals interfere with the synthesis of particular neurotransmitters, and others block a neurotransmitter's access to its receptor molecule (Marwaha and Anderson, 1984). Neuroactive pharmaceuticals commonly have those types of mechanisms of action. Some neurotoxicants are metabolized by neuronal enzymes and produce damaging metabolites. For example, 6-hydroxydopamine is taken up by the re-uptake system at synapses that use dopamine as a neurotransmitter. The metabolism of 6-hydroxydopamine within the synaptic bouton yields hydrogen peroxide and other toxic species, which kill the cell (Graham et al., 1978).

Ion Channels

A nerve impulse depends on the proper functioning of ion-specific channels in the membrane. During the rising phase of the action potential, sodium ions enter the neurons through open sodium channels. During the falling phase, potassium ions leave neurons through open potassium channels. Thus, the internal sodium concentration increases and the internal potassium concentration decreases during an action potential. Although the resulting ionic imbalance is small (around 0.1% for an axon 1 μm in diameter), it must be restored. That is accomplished by the sodium-potassium pump, which extrudes sodium and absorbs potassium. The ion pump is operated by metabolic energy, so substances that inhibit metabolic enzymes cause sodium to accumulate in the neurons and potassium to be lost; that leads to membrane depolarization and, eventually, the loss of excitability. Several kinds of neurotoxic chemicals—including such local anesthetics as lidocaine and such natural toxins as tetrodotoxin and saxitoxin—block the sodium and potassium channels and thereby halt the electric conduction of nerve impulses.

EXAMPLES OF NEUROTOXIC MECHANISMS

In some kinds of neurotoxicity, understanding of the pathogenetic progression from molecular lesion to clinical manifestation is well developed. Pyrethroids cause acute, but reversible, neurotoxic effects. Chronic, but not acute, exposure to *n*-hexane results in degeneration of motor and sensory axons. MPTP, in sufficiently high doses, causes acute degeneration of dopamine neurons, which results in a severe parkinsonian syndrome; however, it is suspected that the progressive reduction in dopamine neurons that occurs with age will result in additional cases of parkinsonism in people exposed to smaller amounts of MPTP (Calne et al., 1985). The possibility of long latency between exposure to a toxicant and clinical manifestation of disease has profound implications both for demonstrating a toxic etiology in degenerative CNS disease and for testing chemicals for neurotoxicity.

Pyrethroids

Pyrethroids are synthetic analogues of pyrethrins, the active substances in the flowers of *Chrysanthemum cinerariaefolium*. The pyrethrum insecticides (made from the dried flowers) had been used extensively until the end of World War II, but a variety of potent synthetic insecticides developed after the war—such as DDT, lindane, dieldrin, parathion, and malathion—made pyrethrum almost obsolete. The pendulum swung in the 1960s, when the persistence of pesticides became a major issue in their use. Many pyrethrin derivatives were then synthesized and tested as insecticides, and some of them have proved extremely useful and safe. The pyrethroid insecticides now in use are characterized generally by potent insecticidal activity, relatively low acute mammalian toxicity, little known chronic mammalian toxicity (including mutagenicity and carcinogenicity, although information is incomplete), and—what is very important—environmental biodegradability.

Among insecticides, pyrethroids and organophosphate-carbamate anticholinesterases have been studied most thoroughly for their mechanisms of action. Pyrethroids exert primarily acute toxicity, so studies have focused on how the signs of

poisoning are produced. The major aim of the studies has been to generate information for improving pyrethroids as insecticides.

Pyrethroids can be classified into two large groups (Figure 2-3). Type I pyrethroids do not contain a cyano group in their molecules and include allethrin, tetramethrin, permethrin, and phenothrin. Type II pyrethroids contain a cyano group at the α-carbon position and include newer compounds, such as deltamethrin, cyphenothrin, cypermethrin, and fenvalerate. The two types of pyrethroids cause somewhat different symptoms of mammalian poisoning. Poisoning with type I pyrethroids is characterized by hyperexcitation, ataxia, convulsions, and eventual paralysis; poisoning with type II pyrethroids, by hypersensitivity, choreoathetosis, tremors, and paralysis. Despite differences in the symptoms, both types of pyrethroids have the same major target site: the sodium channel of nerve membrane, i.e., the channel directly responsible for generating action potentials.

Pyrethroids cause essentially similar signs

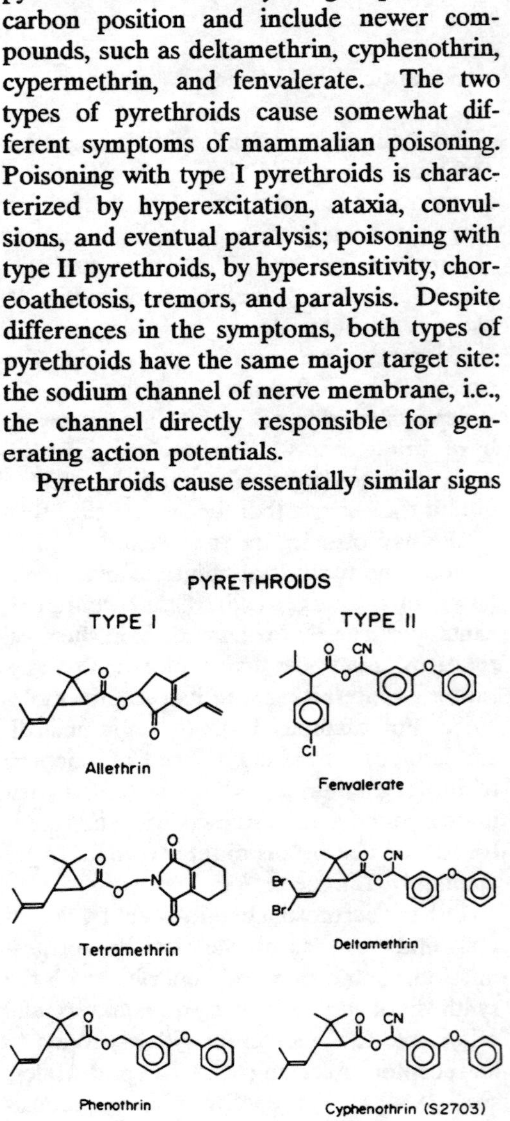

FIGURE 2-3   Structures of type I and type II pyrethroids. Source: Narahashi (1985).

of poisoning in mammals and invertebrate animals. In vitro studies of their mechanism of action on the nervous system have used nerve preparations isolated from insects, crayfish, and squid and cultured mouse neuroblastoma cells (Narahashi, 1971, 1985, 1989; Ruigt, 1984; Wouters and van den Bercken, 1978).

Pyrethroid poisoning is associated with an increase in electric activity of the central and peripheral nervous systems: a single presynaptic stimulation in the presence of pyrethroids causes repetitive excitation of presynaptic nerve and nerve terminals, thereby evoking repetitive postsynaptic discharges. This can be demonstrated by exposing isolated nerve fiber preparations, such as crayfish giant axons and frog myelinated nerve fibers, to low concentrations of pyrethroids (especially type I pyrethroids). Intracellular potential recording experiments reveal that the depolarizing after-potential is gradually increased in amplitude and prolonged after application of pyrethroids, until it finally generates repetitive after-discharges (Lund and Narahashi, 1983).

How the depolarizing after-potential is increased by pyrethroids is best studied by voltage-clamp techniques, whereby membrane ionic currents can be recorded. The sodium current that generates action potentials is markedly prolonged after the preparation is exposed to pyrethroids, and that causes a sustained depolarization after an action potential (Narahashi and Lund, 1980). However, the sodium current thus recorded represents an algebraic sum of currents passing through a large number of open sodium channels. The activity of individual ion channels can be studied by patch-clamp techniques, which allow measurements of ionic currents flowing through individual open channels. Pyrethroids have been found usually long periods (Yamamoto et al., 1983). A sodium channel normally is kept open during a depolarizing step for only a few milliseconds, whereas a sodium channel exposed to pyrethroids can remain open

much longer—even up to several seconds, depending on the pyrethroid in question. Type I and type II pyrethroids have similar effects, but the increase in open time is more pronounced with type II pyrethroids.

Pyrethroids exert another effect on sodium channels. A pyrethroid-exposed sodium channel can be opened at negative membrane potentials near the resting potential. Prolonged opening of many sodium channels near the resting membrane potential leads to the nerve-membrane depolarization observed in the presence of pyrethroids. Membrane depolarization produces several changes in nervous function: depolarization in sensory neurons sends massive discharges to the central nervous system, causing hypersensitivity to external stimuli and paresthesia or a tingling sensation in the facial skin; depolarization of presynaptic terminals increases transmitter release, thereby disturbing synaptic transmission; and depolarization beyond some magnitude blocks nerve conduction and results in paralysis. In addition to the modified electric behavior of the sodium channels, all the more readily detected changes are biologic markers of exposure to pyrethroids.

During the last several years, it has been hypothesized that the γ-aminobutyric acid (GABA) receptor-channel complex, rather than the sodium channels, might be the target site of type II pyrethroids. That was based on several observations, including pyrethroid inhibition of ligand binding to the GABA receptor channel (Lawrence and Casida, 1983). However, recent patch-clamp experiments with rat dorsal root ganglion neurons in culture have unequivocally demonstrated that, whereas sodium channel current undergoes drastic and characteristic prolongation during exposure to the type II pyrethroid deltamethrin, the GABA-induced chloride channel current remains unaffected (Ogata et al., 1988). The GABA receptor-channel system thus plays a negligible role in poisoning with type II pyrethroids.

Pyrethroids thus seem to modify the

gating kinetics of sodium channels, causing prolonged opening of individual channels, which leads to a wide range of toxic signs. Only a very small fraction of sodium channels need to be modified by pyrethroids to cause severe signs of poisoning (Lund and Narahashi, 1982). That is why pyrethroids are such potent insecticides.

Although the mechanism of action on nerve membrane sodium channels is generally the same in both mammals and insects, pyrethroids are less toxic to mammals than to insects. The selective toxicity to insects might be due to at least three major factors: pyrethroids are more effective on nerves at low temperature (Narahashi, 1971), so insects at room temperature would be affected more severely than mammals at 37°C; pyrethroids are detoxified by enzymes, and enzymatic reactions are slower at low temperature; and, because insects are smaller than mammals, pyrethroid molecules are likely to reach the target site before being inactivated metabolically.

In summary, pyrethroids primarily modify gating kinetics of sodium channels, and this action can account for the various symptoms of poisoning in animals. Future studies should be directed toward the molecular mechanisms whereby the pyrethroid molecule interacts with sodium channels, including the identification of the subunit of a channel to which pyrethroids bind.

γ-Diketones

At high concentrations, the simple alkane n-hexane (a glue solvent) can be acutely toxic; e.g., it can cause dizziness and eye, throat, and skin irritations, all of which are nonspecific biologic markers of exposure and effect. Those manifestations, however, are reversible. The progressive weakness and sensory loss that arise from chronic exposure to n-hexane are the major human health hazards that motivated its regulation.

After it was reported that workers exposed to n-hexane at high concentrations developed a motor-sensory peripheral neuropathy (Yamamura, 1969; Herskowitz et al., 1971), Schaumberg and Spencer (1976) showed that rats develop an identical neuropathy after inhaling n-hexane. The finding that humans exposed to a solvent mixture containing methyl butyl butane developed peripheral neuropathy identical with that found after n-hexane exposure (Billmaier et al., 1974) demonstrated the peripheral neuropathy as a biologic marker characteristic of the two chemicals.

Biopsies of humans and careful morphologic studies of exposed rats showed that exposure to n-hexane and exposure to methyl n-butyl ketone each resulted in large neurofilament-filled swellings of axons, with axonal degeneration distal to the swellings developing over time. Thus, a biologic marker was established that was useful across species, as well as for the chemical class. Cavanagh and Bennetts (1981) showed that the neurofilament-filled swellings in the CNS and the PNS uniformly developed along the distal portions of long axons; however, accompanying axonal degeneration was much less common in the CNS (Griffin and Price, 1980). Spencer and Schaumberg (1977a,b) classified the neuropathy as a central-peripheral distal axonopathy and observed that myelinated axons were more vulnerable than nonmyelinated axons. Swelling and distal degeneration were most prominent in myelinated axons of the largest diameter.

The common morphologic findings after n-hexane and methyl n-butyl ketone exposure suggested a metabolic relationship between the two toxicants, which was elucidated by Krasavage et al. (1980) and Couri and Nachtman (1979). Studies disclosed a common metabolite, 2,5-hexanedione (HD), a biologic marker of exposure that itself caused an identical neuropathy. The Krasavage group coined the term "γ-diketone" neuropathy to emphasize the importance of the gamma spacing of the two ketone groups (separated by two carbons; see Figure 2-4). γ-Diketones and γ-diketone precursors produce a specific neurotoxic response not shared by diketones with α, β, or γ spacing (with zero, one, or three intervening carbons, respectively) (Spencer et al., 1978; Katz et al., 1980). However, 2,4-pentanedione, an α-diketone, produces a different form of neurotoxicity (Misumi and Nagano, 1984). The seminal observation regarding the significance of gamma spacing began the process of understanding the neuropathy at the molecular level.

The search for the fundamental mechanism focused on two concepts: inhibition of energy metabolism (Spencer et al., 1979) and neurofilament cross-linking (Graham, 1980). The key observation, however, was that lysyl amino groups of proteins are the primary target of γ-diketones and that pyrroles are the product of the reaction (DeCaprio et al., 1982; Graham et al., 1982).

A number of studies have tested the hypothesis that pyrrole formation is the initial step in the sequence that leads to neurofilament accumulation. Anthony et al. (1983a,b) showed that 3,4-dimethyl-2,5-hexanedione (DMHD) was a γ-diketone that

TOXIC RESPONSES OF THE NERVOUS SYSTEM

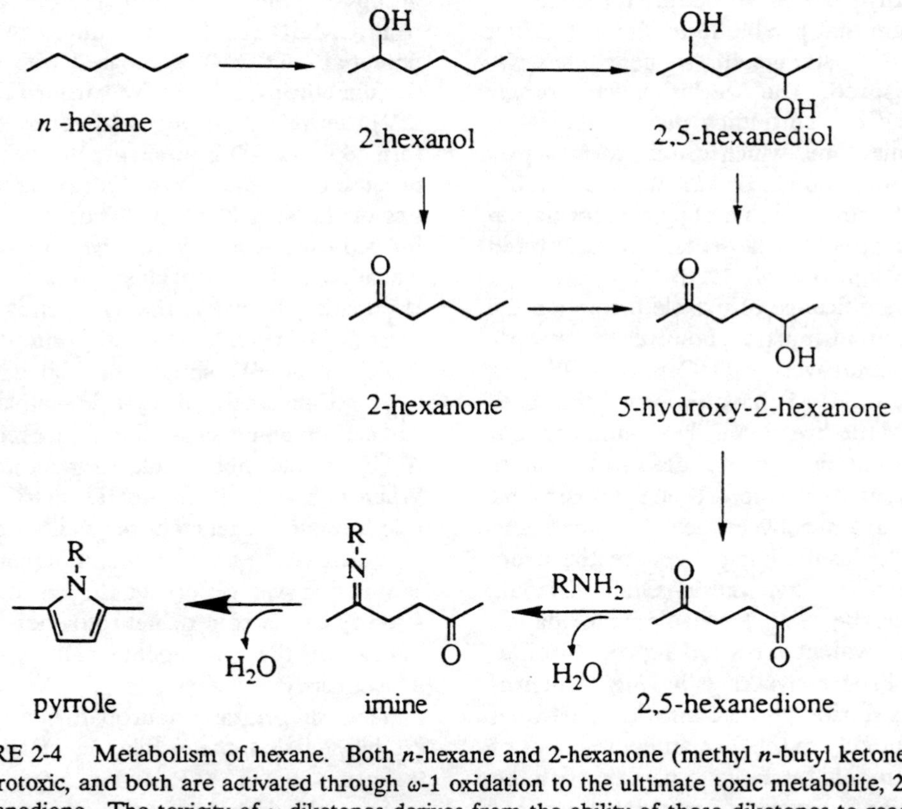

FIGURE 2-4   Metabolism of hexane. Both *n*-hexane and 2-hexanone (methyl *n*-butyl ketone) are neurotoxic, and both are activated through ω-1 oxidation to the ultimate toxic metabolite, 2,5-hexanedione. The toxicity of γ-diketones derives from the ability of these diketones to react with protein amino groups (RNH$_2$) to form an amine in an initial reversible step, and then to cyclize irreversibly to form a pyrrole. Source: Anthony and Graham (1991). Reprinted with permission from Amdur et al. (1991).

formed pyrroles more rapidly than HD and was correspondingly more neurotoxic. Larger substituents—such as ethyl, isopropyl, and phenyl groups on the 3 and 4 carbons—impeded pyrrole formation by steric hindrance (Szakal-Quin et al., 1986) and were not neurotoxic (Genter et al., 1987). Thus, although the gamma spacing is critical, not all γ-diketones are neurotoxic. Genter et al. (1987) clarified the role of pyrrole formation with the observation that the *dl* diastereomer of DMHD forms pyrroles more readily and is more toxic than the *meso* diastereomer. Given that *dl*-[^{14}C]DMHD and *meso*-[^{14}C]DMHD disperse comparably in the nervous system, as shown by Rosenberg et al. (1987), and that the tetramethylpyrroles formed are the same, the conclusion that pyrrole formation is the first step in the pathogenetic sequence was virtually ensured. The conclusion was strongly supported by the finding that 3,3-dimethyl-2,5-hexanedione, which cannot form a pyrrole, is not neurotoxic (Sayre et al., 1986) and by the reduced rate of pyrrole formation and neurotoxicity of deuterium-substituted HD (DeCaprio et al., 1988).

The significance of pyrrole formation for neurofilament function, however, has been a point of controversy. DeCaprio (1985) and Sayre et al. (1985) have proposed that conversion of the hydrophilic lysyl amino groups to hydrophobic pyrrole derivatives alters neurofilament transport through impairment of interactions between neurofilaments and other cytoplasmic components in the axon. The alternative hypothesis (Graham et al., 1982) has been that pyrrole autoxidation leads to covalent cross-linking of neurofilaments. Progressive cross-linking of neurofilaments during chronic intoxication would lead eventually to the formation of masses of neurofilaments too large to pass through the constrictions in axonal diameter that occur at every node of Ranvier. Indeed, the greater vulnerability of large-diameter myelinated axons is apparently explained by the greater proportional reduction in diameter at their

nodes. Thus, continued transport up to the point of occlusion could account for the large neurofilament-filled swellings of the axon that most often occur proximal to nodes of Ranvier (Spencer et al., 1977a; Jones and Cavanaugh, 1983).

To test whether pyrrole derivatization itself is sufficient to account for the neurofilament-filled swelling or whether neurofilament cross-linking is necessary, St. Clair et al. (1988) used another γ-diketone, 3-acetyl-2,5-hexanedione (AcHD). The acetyl group of AcHD is small enough not to hinder pyrrole formation; in fact, the rate is nearly as great as that for *dl*-DMHD and greater than that for *meso*-DMHD. However, whereas the electron-donating methyl groups of the tetramethylpyrrole derived from DMHD result in a reduced oxidation potential (362 mV), compared with that of the dimethylpyrrole derived from HD (620 mV), the oxidation potential of the pyrrole formed by AcHD is much greater (975 mV), because of the electron-withdrawing properties of the acetyl group. Thus, the pyrrole formed from AcHD is relatively resistant to oxidation and so provides a probe for discriminating between the competing hypotheses. If pyrrole derivatization itself is sufficient, AcHD should be highly neurotoxic. Conversely, if pyrrole autoxidation and neurofilament cross-linking are required, AcHD would not result in neurotoxicity. When rats were given AcHD, massive pyrrole formation occurred, but neither clinical neurotoxicity nor neurofilament accumulation was observed; the results point strongly to the role of neurofilament cross-linking in the pathogenesis of *n*-hexane neurotoxicity.

Thus, in *n*-hexane neuropathy, it appears established that the γ-diketone HD is the toxic metabolite. Reaction with amino groups yields pyrrole adducts, which undergo autoxidation that results in covalent cross-linking of proteins. It is the extreme stability of the neurofilaments that results in the localization of toxic injury in the nervous system, and it is the particular anatomy of the myelinated axon at nodes of Ranvier that makes these axons most vulnerable.

MPTP

A particularly unfortunate byproduct of the so-called designer drugs produced in the 1980s was 1-methyl-4-phenyl-1,2,3,6-tetrahydropyridine (MPTP) (Figure 2-5). Instead of the desired opiate-like characteristics of the intended meperidine derivative (which was to be sold illicitly as "synthetic heroin"), a synthetic drug that contained MPTP caused a profound, acute-onset neurologic disorder with extreme muscular rigidity and tremor that was indistinguishable from idiopathic Parkinson's disease (Langston et al., 1983). Animal studies showed that MPTP itself is not neurotoxic, but requires metabolism by monoamine oxidase B, a process that appears to take place in astrocytes, a type of glial cell (Jenner, 1989). The product of MPTP oxidation is the pyridinium ion MPP+ (Figure 2-5); this putative toxic metabolite of MPTP leaves the astrocytes and is preferentially concentrated by catecholaminergic neurons via their own catecholamine uptake system. Whether the mechanism of MPP+ cytotoxicity is related to oxidation-reduction cycling, interference with mitochondrial energy production, or some other process, the end result is cell death, particularly death of the dopaminergic neurons of the substantia nigra (Langston and Irwin, l986). The marked depletion of dopamine seen in the brain of MPTP-treated

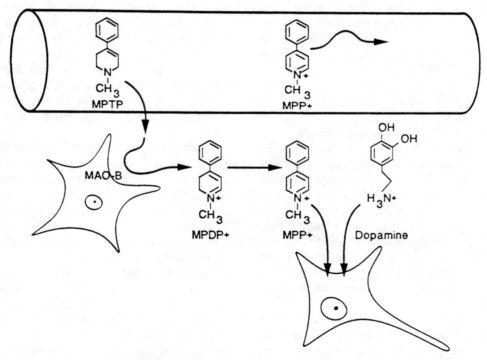

FIGURE 2-5   Diagram of MPTP toxicity. MPP$^+$, either formed elsewhere in the body following exposure to MPTP or injected directly, is unable to cross the blood-brain barrier. In contrast, MPTP gains access and is oxidized in situ to MPDP$^+$ and MPP$^+$. The same transport system that carries dopamine into the dopaminergic neurons also transports the cytotoxic MPP$^+$. Source: Anthony and Graham (1991). Reprinted with permission from Amdur et al. (1991).

animals is also the central neuropathologic feature of Parkinson's disease in humans.

Not all people exposed to MPTP have developed parkinsonism, ostensibly because the dose administered was sometimes insufficient to cause death of a large enough proportion of dopaminergic neurons to produce the 80% or greater depletion of dopamine in the brain that is required to produce a clinical syndrome (Hornykiewicz, 1986). Recently, a number of exposed people have been studied with positron-emission tomographic (PET) scanning (Calne et al., 1985). Through the use of positron-emitting fluorodopa, which is sequestered in the striatum (where the axonal projections of the substantia nigra form synapses), a deficiency of striatal dopamine was demonstrated. Although the exposed people were asymptomatic, the loss of dopaminergic neurons with normal aging (McGeer et al., 1977) is likely to reduce the population of these cells, and thus brain dopamine concentrations, below the 80% required for normal function; at that time, the signs and symptoms of Parkinson's disease would be manifested (Calne and Langston, 1983).

This is a critically important concept: An acute toxic injury can be separated from its clinical manifestations by years. If MPTP-exposed people who are now asymptomatic (and there might be over 500 such cases) eventually develop parkinsonism, it will open up the possibility that some patients who develop progressive degenerative CNS disease later in life (such as Parkinson's disease, Alzheimer's disease, and amyotrophic lateral sclerosis) might have suffered neuronal injury from a toxicant years earlier. If that turns out to have been the case, tests for neurotoxicity will have to take into account the possibility that effects develop after a long latency.

SUMMARY

The nervous system has several unique aspects, e.g., poor regenerative capacity, unusual anatomy, specialized metabolic requirements and subcellular systems, and a wide variety of neurotransmitter and neuromodulator systems. As the discussions of pyrethroids, γ-diketones, and MPTP illustrate, the complexity of the nervous system and its tremendous structural and functional heterogeneity might make it particularly susceptible to disruption by toxic agents, and neuronal dysfunction can have impacts on a number of organ systems. Even minor changes in the structure and function of the nervous system can have profound effects on neurologic, behavioral, and other body functions.

Many foreign substances can alter the normal activity of the nervous system via a variety of mechanisms. Some produce immediate short-term effects; others cause neurotoxic effects only after long-term, repeated exposures. Some substances can cause permanent damage after a single exposure; others can cause subtle and reversible damage. This diversity in nervous system response increases the challenge to toxicologists to develop biologically appropriate and empirically reliable test systems.

Considerable progress has been made in recent decades in understanding the nervous system on an integrated and comprehensive basis. Some mechanisms underlying behavior, such as sensorimotor function and information-processing, are being elucidated from the translation of molecular and cellular activities into organ-level events and from the translation of organ-level events into whole-organism behavior. Only by understanding the basic mechanisms involved in neurobiologic processes can we develop adequate neurotoxicity test methods. Our present understanding of some expressions of nervous system function permits such an integrated approach—e.g., much is known about the roles of molecules, cells, organs, and the whole organism in visual processing and proprioceptive control of movement. For most other neurologic functions, although gaps in our knowledge still restrict integrative analyses, we can examine their components at specific levels of integration-with in vivo and in vitro approaches. An understanding of the nervous system guides the search for useful biologic markers, as is discussed in Chapter 3. This information is developed further in Chapter 4 to assess available in vivo and in vitro testing methods and address the need for efficient means for identifying neurotoxic substances.

3

Biologic Markers in Neurotoxicology

The purpose of this chapter is to form a bridge between the issues of nervous system structure and function that are the principal subjects of contemporary neuroscience (Chapter 2) and the goals and emerging methods of quantitative risk assessment (Chapter 6). Identification of biologic markers and understanding of their role in cell response and damage can provide information that will be essential for predicting, and therefore preventing, diseases and disability. This chapter applies to the nervous system the important conceptual framework developed recently by the National Research Council's Committee on Biologic Markers and applied in pulmonary toxicology (NRC, 1989a), reproductive and neurodevelopmental toxicology (NRC, 1989b), and immunotoxicology (NRC, 1991a). We believe that the concepts and definitions of the Committee on Biologic Markers constitute a valuable tool that can be applied to neurotoxicology, as described in the following sections.

CONCEPTS AND DEFINITIONS

Biologic markers, broadly defined, are indicators of change or variation in cellular or biochemical components or processes,

structures, or functions that are measurable in organisms or samples from those organisms. Understanding the processes, structures, and functions of the nervous system is the starting point for identifying biologic markers. Current interest in the biologic markers of the nervous system stems from a desire to identify the early stages of system impairment and damage and to understand the dynamics of exposure and the basic mechanisms of response to toxic exposure.

There is growing interest in the use of biologic markers in clinical medicine, epidemiology, toxicology, and related biomedical fields to study the human neurologic effects of exposures to environmental toxicants (NRC, 1989a,b). Clinical medicine uses markers to allow earlier detection and treatment of disease; epidemiology uses them as indicators of internal dose or of health effects and for surveillance; and toxicology uses them to help to determine underlying mechanisms of diseases, develop better estimates of dose-response relationships, and improve the technical bases for assessing risks associated with low levels of exposure (see Chapter 6).

The rapid growth of new molecular and biochemical tools in the practice of medicine has resulted in the development of markers for understanding disease, predicting outcome, and directing treatment. Over the next few years, many of the tools will be used to estimate chemical exposure. Even now, many diseases are defined, not by overt signs and symptoms, but by the detection of biologic markers at the subcellular, molecular, or functional level after exposure to a toxic agent. The identification, validation, and use of markers in neurobiology and medicine depend fundamentally on increased understanding of mechanisms of action and of molecular and biochemical processes. Biologic markers are allowing investigators to gain a better understanding of chemical-receptor interactions. As noted in Chapter 2, for example, some toxicants interfere with ion channels and thus enhance or suppress impulse conduction along nerves. Tetrodotoxin, a puffer fish poison, blocks the sodium ion channel, thereby impairing impulse conduction. However, the insecticidal pyrethroids keep the sodium ion channels open too long, thereby making the nerve hyperactive.

As we use the term here, markers can be signals or indicators of normal physiology, forerunners of nervous system damage, or clinical signs and symptoms of neurologic disease. A specific biologic marker can serve several purposes and is best characterized by the use to which it is put in a particular context. Markers can indicate susceptibility, exposure to an exogenous agent, internal dose, biologically effective dose (dose at a receptor site), early biologic effect, structural or functional alteration, physiologic status, system damage, or disease. Thus, the choice of a marker and its interpretation depend on the purpose of its use, and its intended use depends on characteristics peculiar to a particular exogenous agent, to an individual organism, or to a particular target organ or tissue (see Table 3-1). If the goal is prevention, the major emphasis will be on monitoring markers that

TABLE 3-1 Examples of Characteristics of Exogenous Agents, Organisms, or Targets That Influence Choice of Biologic Marker

Agent-specific characteristics	Organ- or tissue-specific characteristics
Physiochemical properties	Blood flow
Interactions	Membrane permeability
Routes of exposure	Transport
Duration of exposure	Receptors
Exposure concentration	Function
Pattern of exposure	Homeostasis
Metabolism	Structure
Activation	Physiologic state
Detoxification	
Organism-specific characteristics	
Species	
Age	
Sex	
Physiologic state	
Pharmacokinetic characteristics	
Genetic factors	
Lifestyle factors	

Source: NRC (1989b).

identify biologic changes in the nervous system that are predictive of health impairment or overt disease, such as the various biochemicals, metabolites, responses, and characteristics listed in Table 3-2.

Markers represent points in a continuum between health and disease. How they are applied to the neurologic sciences might change as our knowledge of the fundamental processes of disease of the nervous system increases. What are perceived at first to be early signals of exposure and of a system's response to a toxicant could come to be considered health impairments themselves, because of their strong predictive relationship to disease. By the time early changes (markers) are detected, it might be difficult to prevent neuropathy from occurring in response to exposure to a toxicant whose effects are progressive and irreversible. But many neurotoxicants produce damage that is reversible if exposure is stopped (e.g., agents that cause CNS depression). Thus, biologic markers related to the nervous system can be valuable in prevention, in estimation of exposure, in early detection of damage, and in early treatment of disease.

The Committee on Biologic Markers found it useful to define three general categories of biologic markers: those of exposure to chemical or physical agents, those of effects of exposure, and those of susceptibility to the effects of exposure (NRC, 1989a,b). In the context of the nervous system, a biologic marker of exposure is an exogenous substance or its metabolite or the product of an interaction between a substance and some target molecule or other nervous system receptor. A biologic marker of effect is a detectable alteration of an endogenous component within the nervous system that, depending on magnitude, can be recognized as a potential or established health impairment or disease. A biologic marker of susceptibility is an indicator of an inherent or acquired limitation of an organism's nervous system's ability to respond to the challenge of exposure to a specific

TABLE 3-2 Selected Markers of Neurotoxicity in Nervous System

General neuronal measures
  Cell number
  Tetanus-toxin binding
  Neurofilament protein
  Neuronal structure

General glial measures
  Glial fibrillary acidic protein
  Oligodendrocyte probe

Transmitter systems
  Amino Acid
    Excitatory
    Inhibitory
  Cholinergic
    Choline acetyltransferase
    Muscarinic and nicotinic receptors
  Aminergic
    Norepinephrine
    Serotonin
    Dopamine
  Peptidergic
    Vasoactive intestinal peptide
    Substance P
    Enkephalin

Cell biologic responses
Second messengers
    Cyclic nucleotide
  Phosphorylation

Calcium-dependent transmitter release

Voltage-dependent NA^+ or Ca^{2+} uptake

toxic substance. Organisms might be particularly susceptible to a toxic substance for several reasons, including a genetic predisposition, existing disease, unique metabolic characteristics, or even the consequences of increased stress (see discussion on markers of susceptibility).

Biologic Markers of Exposure

External exposure is the amount or concentration of a substance in the environment of an organism; internal dose is the amount that is transferred or absorbed into the organism (see Figure 3-1). Biologically effective dose, in general terms, is the internal dose that is quantitatively correlated with an identifiable biologic effect; however, it is more precisely considered to be the amount of a substance that has interacted with a critical cellular or tissue receptor or target where the biologic effect is initiated. Because many such receptor sites in the nervous system are not known or not accessible, it is often necessary to use a surrogate site. For example, the neurofilament cross-linking that follows exposure to γ-diketones is accompanied by dimerization of erythrocyte spectrin; blood samples can reveal that a biologically effective dose has been received by the organism (St. Clair et al., 1988) in that these compounds penetrate to the nervous system sites in proportion to concentrations in blood samples.

Markers of exposure can be based on steady-state or pharmacokinetic measures, such as peak circulating concentration, cumulative dose, or plasma half-life. Individual variations in physiologic characteristics—such as sex, age, blood flow, membrane permeability, and respiratory rate—can affect the absorption and distribution of a chemical and its metabolites (Doull, 1980; NRC, 1986). Age and health status, such as disease, can alter the respiratory rate and thus the pulmonary dose of a toxicant (NRC, 1989a). Exposure concentration, size of delivered dose, and dose rate also affect internal dose. When absorption capacities are exceeded, alternate pathways of clearance come into play. High vapor concentrations (e.g., high concentrations of organic solvents) might be exhaled, rather than absorbed. Species and individual differences in metabolism can drastically alter internal doses of reactive metabolites.

The internal dose of a substance can vary with route of exposure, chemical species, and physical form. To make qualitative or quantitative estimates of exposure with biologic markers, the concentration, duration and pattern of exposure, and physicochemical nature of a toxicant must be considered in the selection of an appropriate marker of exposure (Gibaldi and Perrier, 1982). Other environmental factors, such as temperature, can affect exposure by changing amounts of water consumption and thus of waterborne pollutants ingested. Diet alters intestinal motility and gastric emptying time, as well as the transport of specific substances across barriers in the gastrointestinal system. The

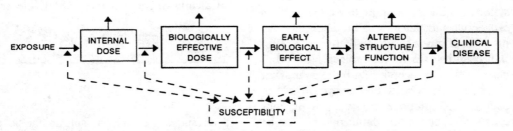

Figure 3-1. Simplified classification of biologic markers (indicated by boxes). Solid lines indicate progression, if it occurs, to the next class of marker. Dashed lines indicate that individual susceptibility influences the rates of progression, as do other variables described in the text. Biologic markers represent a continuum of changes, and the classification of change might not always be distinct. Source: NRC Committee on Biologic Markers, 1987.

formulation or type of mixture that contains a toxicant can affect its transport across protective barriers and membranes in the nervous system.

The presence of active mechanisms of transport into an organ or tissue and the density of nervous system receptor sites can all influence internal dose and biologically effective dose. Metabolism critically affects biologically effective doses of many compounds; the tissue distribution of metabolizing enzymes is an important determinant of effective dose. Many nervous system toxicants act by interfering with neurotransmitter-receptor interactions. Thus, the interpretation of dosimetric data involves an understanding of the role of the receptor in overall cell, organ, or organism function, of coexisting or pre-existing stresses on the organism, and of the existence and availability of repair or compensation during and after exposure (Doull, 1980). These interactive aspects of exposure and response are especially important for the nervous system, which participates in the physiologic regulation of other body systems.

An example of a marker of exposure is the inhibition of blood acetylcholinesterase by organophosphorus pesticides. However, although moderate levels of inhibition—i.e., up to 30%—are not associated with physiologic effects (NRC, 1986), inhibition of blood acetylcholinesterase greater than 30% can be used as a marker of both exposure and effect. As another example, urinary concentrations of 2,5-hexanedione have been suggested as markers of exposure to n-hexane.

Biologic Markers of Effect

For purposes of environmental health research, biologic markers of effects in the nervous system of an organism after exposure to a toxicant are considered in the context of their relationship to health status

—normal health, early health impairment and damage, or overt disease (see Figure 3-1). In that context, an effect is defined as any of the following:

- An alteration in a tissue or organ.
- An early event in a biologic process that is predictive of the development of a health impairment.
- A change consistent with early damage to an organ or tissue.
- A health impairment or clinically recognized disease.
- A response peripheral or parallel to a disease process, but correlated with it and thus usable in predicting development of a health impairment.

Thus, a biologic marker of a nervous system effect can be any qualitative or quantitative change that is predictive of system impairment or damage resulting from exposure to a substance, e.g., in brainwave recordings, in release of a neurotransmitter (such as dopamine), or in the presence of amyloid in Alzheimer's disease or of neuron-specific enolase in some CNS neoplasms. Markers of early biologic effects include alterations in the functions of target tissues of the nervous system after exposure. As early-warning signals, such markers can be useful dosimeters to guide intervention aimed at reducing or preventing further exposure. Such early-warning signals might also be observed in organs or tissues other than the sites that are critical for toxic action, in that effects on the nervous system can themselves affect the functioning of other organ systems—e.g., caffeine can stimulate the CNS, which can then affect the heart and kidneys.

A tissue affected by a toxicant might exhibit altered function even if the exposed organism has no overt manifestations. Such alteration can in some cases be determined by testing, particularly with the new biochemical probes (e.g., molecular adducts) or imaging techniques (e.g., magnetic resonance imaging). Biologic markers of functional alterations are most useful if related to specific functions—e.g., acetylcholinesterase for a cholinergic response and epinephrine for a catecholaminergic response.

If the internal dose of a toxicant is great enough, disease will develop, because the biologically effective dose will be sufficient to affect some function irreversibly or produce an effect for a substantial period, as in the case of ethanol. Disease that occurs soon after exposure might be easy to link directly to the toxicant. Disease that occurs long after exposure might be difficult to relate to a specific toxicant, as is postulated for some cases of Alzheimer's disease and other neurodegenerative disorders (Calne et al., 1986), unless the findings are pathognomonic, i.e., are relatively specific to a particular type of exposure (such as γ-diketones and methyl n-butylketone) or are rare in unexposed persons (such as distal axonopathy).

The transition to overt disease can depend on properties of the toxicant, the nature of exposure, the disease process itself, or individual susceptibilities. Because people respond differently to toxicants, it is not surprising that only some members of a population similarly exposed to a given environmental agent will develop a given disease. That is especially true of the nervous system, in which large observed variations in population response to chemical toxicants have been attributed to differences in metabolism, pharmacokinetics, protective mechanisms and barriers, and system complexity.

Although scientists tend to divide biologic markers into groups, there is a continuum between health and disease, and advances in toxicology have demonstrated a continuum between dose and response. Biologic markers are best divided operationally, depending on how they are assessed and how they can be used, but the divisions should not be interpreted to imply mechanistic distinctions (Figure 3-1).

Biologic Markers of Susceptibility

Some biologic markers indicate individual or population factors that can affect response to environmental agents. Those factors are independent of whether exposure has occurred, although exposure sometimes increases susceptibility to the effects of later exposures (e.g., in sensitization to formaldehyde). An intrinsic characteristic or pre-existing disease state that increases the internal dose or the biologically effective dose or that amplifies the effect at the target tissue can be a biologic marker of increased susceptibility (NIEHS, 1985; Omenn, 1986). Such markers can include inborn differences in metabolism, variations in immunoglobulin concentrations, low organ reserve capacity, or other identifiable genetically or environmentally induced factors that influence absorption, metabolism, detoxification, and effect of environmental agents. These types of biologic markers are not extensively considered in this report.

VALIDATION OF BIOLOGIC MARKERS

To validate the use of a biologic measurement as a marker, it is necessary to understand the relationship between the marker and the event or condition of interest, e.g., between muscular weakness and exposure to acrylamide. Sensitivity and specificity are critical components of validity (MacMahon and Pugh, 1970). Sensitivity is the ability of a marker to identify correctly those who have been exposed or have a disease or condition of interest; specificity is the ability to identify correctly those who have not been exposed or do not have the disease or condition.

Animal models are useful for understanding the basic mechanistic processes of the expression of markers and the relationships among exposure, early effects, and disease and thus are useful in validating biologic markers (see Chapter 4). If a disease can be satisfactorily induced in experimental animals (e.g., axonopathy in rats exposed to *n*-hexane), then potential biologic markers for predicting eventual disease can be explored, and early indicators of the disease might be identified for use in human studies (e.g., peripheral neuropathy). A major goal of research with biologic markers is to develop markers that reliably indicate an early stage in the development of a disease process in humans when effective intervention is still possible. That is especially important for the nervous system, where repair is limited and early intervention might offer the only chance of recovery.

One of the most serious challenges for the use of animal models in neurotoxicology is that many expressions in the human central nervous system are quite different from those which can be assessed directly in animal models. Moreover, even for events that appear to be similar, humans might differ from specific experimental animals in their functional reserve capability. Therefore, the degree of impairment required to cause a particular functional deficit could differ for apparently equivalent levels of damage. Use of appropriate biologic markers for changes in neurologic functions that are similar between animals and humans could help determine functionally equivalent amounts of damage. For example, changes in neurotransmitter turnover might provide a basis for interspecies comparability. This application of markers has been useful in neuropharmacology and should find increased application in the development of neurotoxicologic markers. Until recently, such markers were not directly measurable in humans. However, new scanning techniques of positron-emission tomography and

magnetic resonance imaging can provide information in vivo on the neurochemistry of humans and experimental animals.

The principal purpose of markers in health research is to identify exposed people, so that risk can be estimated and disease prevented. Particularly critical for marker validation is the strength of biologic plausibility that supports an association between a change in a specified signal (a marker) and the occurrence of a specific exposure or of a change in a specific effect.

Validation of a specific marker depends on its expected use in monitored populations. Biologic markers observed well before the onset of disease might have low predictive value for the disease itself, but function acceptably as criteria for defining exposed populations and thus be useful for long-term followup. For example, heroin addiction in California might be a useful clue of potential exposure to the neurotoxicant MPTP for epidemiologic research, even though its relationship to a parkinsonian syndrome is indirect. Conversely, an effect marker that is expressed long after exposure could be of relatively little use in exposure assessment, but very important in predicting progression of disease or calculating risk. For example, neurobehavioral manifestation of prenatal exposure to methylmercury can persist well into adulthood.

Quality assurance and quality control are fundamental to the objective development and application of accurate and verifiable biologic markers. The objective of laboratory quality-assurance practices is to ensure that findings reported by one laboratory are in fact verifiable and within acceptable limits of measurement error, that they accurately indicate the concentrations or presence of the biologic changes reported to have been found, and that they are objective and free from sources of bias introduced through the analytic process. General issues of quality assurance and quality control have been addressed by documents produced by the

Food and Drug Administration (FDA), the U.S. Environmental Protection Agency, the Organization for Economic Cooperation and Development, and other regulatory organizations. FDA developed a set of guidelines known as good laboratory practices, or GLPs (FDA, 1988), which are now incorporated into the standard procedures of most testing and analytic laboratories for well-established assays. GLPs are intended to reduce the chance of contamination (particularly important in the measurement of biologic markers of exposure) or of changes in biologic variables introduced by sample storage, processing, or measurement (Zeisler et al., 1983). The application of GLPs to analysis of biologic samples, especially human tissue, has been reviewed by operational units of the Centers for Disease Control, the National Institute of Standards and Technology (formerly named the National Bureau of Standards), and various clinical laboratories (ACS, 1980; NCCLS, 1981, 1985). Because most of the biologic markers discussed in this report are still at a research stage, standardized GLP practices regarding them remain to be established for laboratories.

USE OF BIOLOGIC MARKERS IN RISK ASSESSMENT

Cellular and molecular markers can serve as powerful new tools for the assessment of risks associated with exposure to environmental toxicants. Markers that indicate the delivery of an internal dose or a biologically effective dose or the induction of a disease process can be useful in hazard identification—i.e., in the qualitative step by which an environmental agent is causally associated with an adverse effect. Biologic markers can also be used to determine dose-response relationships, especially at the low doses relevant to exposure to many environmental chemicals. In addition, they can be used to quantify actual exposures.

Biologic markers in environmental-health studies offer an opportunity to determine the shape of the lower end of the dose-response curve in humans—an opportunity that has not yet been used extensively in standard human epidemiologic or laboratory animal studies. Nevertheless, it is likely to remain uncommon to have information on concentrations of a toxicant at the site of action within the nervous system.

Many neurotoxic effects can be studied in experiments conducted with doses at which enzymatic detoxification or activation and other biochemical reactions along the causal chain are not saturated—i.e., at which the dose-response relationship is still linear. When dose-response relationships are not linear, measures of dose of active forms of toxicants at relevant sites in the nervous system are important for understanding those relationships. In addition, the use of biologic markers can help to sort out potential nonlinearities with respect to dose and can facilitate approaches to four problems critical in predicting neurotoxicity:

• Differences in sensitivity of the individual to neurotoxicants at different stages of life. (The previous report on biologic markers in reproductive toxicology contains an extensive discussion of neurologic markers of developmental effects [NRC, 1989b].)

• Protective mechanisms and barriers that vary with respect to developmental age, location in the nervous system, and degree of metabolic integrity.

• Repair and compensation processes for types of damage that appear to be at least partly reversible under some conditions of exposure.

• Differences in pharmacokinetics (absorption, distribution, metabolism, and excretion) between chemicals and between animals and humans.

It has long been recognized that sensitivity to adverse effects during development can differ dramatically over short periods—especially during gestation. Developmental signals might be present over short periods at the intensity required to induce axons to find their targets and for other key physiologic and biochemical events to occur; these signals might be disrupted by toxic chemicals. Understanding the physiologic basis of temporal differences in sensitivity will be a key to determining over what periods exposures should be aggregated for best experimental and epidemiologic quantification of risks. For example, Marsh et al. (1987) have shown the relationship between the incidence of a variety of fetal methylmercury effects and the time of maximal exposures during gestation, as indicated by sequential mercury concentrations in the hair of the mothers in the Iraqi mass-poisoning incident (see Chapter 1).

It is important to note that, because of the metabolism and the pharmacodynamics of chemical toxicants, there is a continuing need for the development of better and more accurate markers of internal dose that can be used to estimate biologically effective doses. That refinement of dose-response modeling is fundamental to risk assessment.

Neurons do not regenerate in adult life, and some important neurologic conditions (including Alzheimer's disease, Parkinson's disease, and amyotrophic lateral sclerosis) are associated with progressive cumulative loss of specific types of neurons. However, considerable numbers of cells can be lost before clinical signs of impairment are detectable. Thus, exposure to agents causing such loss might proceed without apparent effect for years. Even after symptoms appear, identifying the potentially relevant exposures among those of a lifetime is nearly impossible.

Epidemiologic research on neurodegenerative diseases would be greatly facilitated if either or both of the following types of biologic markers were available:

• *Measures of the past accumulation of relevant damage in people who have not yet developed clinical illness*. For example, can our new tools for imaging the brain be applied to develop a census of the relevant cells left in people of different ages with different past exposures? Imaging methods are being applied to studies of people exposed to doses of MPTP that have not yet produced clinical effects. It is analogous to the use of chest x-rays, FEV_1 (forced expiratory volume in 1 second), and FVC (forced vital capacity) to assess the accumulation of chronic respiratory damage (NRC, 1989a).

• *Measures of the current rate of loss of the relevant cells*. When cells die, do they release measurable amounts of a distinctive form of an enzyme or some other cellular component that could be used as a biologic marker? Distinctive patterns of electric activity in discrete brain regions can be monitored, as can the integrity of neurochemical systems, e.g., receptor binding, synthetic enzyme activity, and glucose utilization.

SUMMARY

In toxicology and epidemiology, biologic markers have several major advantages as means of exploring the relationship between exposures and effects:

• A more complete scientific understanding than a simple input-output analysis based on external dose and the incidence of disease, by virtue of the incorporation of more relevant information about critical events and molecular mechanisms.
• The eventual prospect of improving mechanism-based estimations of risk.
• Improved sensitivity, specificity, and predictive value of detection and quantification of adverse effects at low dose and early in exposure at the cellular or subcellular level.
• The prospect of confirming exposure-related effects and narrowing the search for exposures.
• A better understanding of unique susceptibilities in various population groups.
• A better understanding of disease processes in the nervous system.

Neuroscience and neurotoxicology can be enhanced by directing more research attention to quantitative questions—"How much?" "How fast?" "According to what quantitative functional relationship?"—among the series of causal intermediate processes involved. The role of biologic markers is to measure the steps of disease processes at as many different points as is feasible and to aid the development of means to stop and reverse the processes.

4

Testing for Neurotoxicity

Diseases of environmental origin result from exposures to synthetic and naturally occurring chemical toxicants encountered in the environment, ingested with foods, or administered as pharmaceutical agents. They are, by definition, preventable: they can be prevented by eliminating or reducing exposures to toxicants. The fundamental purpose of testing chemical substances for neurotoxicity is to prevent disease by identifying toxic hazards before humans are exposed. That approach to disease prevention is termed "primary prevention." In contrast, "secondary prevention" consists of the early detection of disease or dysfunction in exposed persons and populations followed by prevention of additional exposure. (Secondary prevention of neurotoxic effects in humans is discussed in Chapter 5.)

In the most effective approach to primary prevention of neurotoxic disease of environmental origin, a potential hazard is identified through premarket testing of new chemicals before they are released into commerce and the environment. Identifying potential neurotoxicity caused by the use of illicit substances of abuse or by the consumption of foods that contain naturally occurring toxins is less likely. Disease is prevented by restricting or banning the use of chemicals found to be neurotoxic or by instituting engineering controls and imposing the use of protective devices at points of environmental release.

Each year, 1,200-1,500 new substances are considered for premarket review by the Environmental Protection Agency (EPA) (Reiter, 1980), and several hundred compounds are added to the 70,000 distinct chemicals and the more than 4 million mixtures, formulations, and blends already in commerce. The proportion of the new chemicals that could be neurotoxic if exposure were sufficient is not known (NRC, 1984) and cannot be estimated on the basis of existing information (see Chapter 1). However, of the 588 chemicals used in substantial quantities by American industry in 1982 and judged to be of toxicologic importance by the American Conference of Governmental Industrial Hygienists (ACGIH), 28% were recognized to have adverse effects on the nervous system; information on the effects was part of the basis of the exposure limits recommended by ACGIH (Anger, 1984).

Given the absence of data on neurotoxicity of most chemicals, particularly industrial chemicals, it is clear that comprehensive primary prevention would require an extensive program of toxicologic evaluation. EPA has regulatory mechanisms to screen chemicals coming into commerce, but overexposures continue to occur, and episodes of neurotoxic illness have been induced by chemicals, such as Lucel-7 (Horan et al., 1985), that have slipped through the regulatory net. Serious side effects of pharmaceutical agents also continue to surface, such as the 3 million cases of tardive dyskinesia that developed in patients on chronic regimens of antipsychotic drugs (Sterman and Schaumburg, 1980). It is now possible to identify only a small fraction of neurotoxicants solely on the basis of chemical structure through analysis of structure-activity relationships (SARs), so in vivo and in vitro tests will be needed for premarket evaluation until greater understanding of SARs permits them to be used with confidence.

If neurotoxic disease is to be prevented, public policy must be formulated as though all chemicals are potential neurotoxicants; a chemical cannot be regarded as free of neurotoxicity merely because data on its toxicity are lacking. Prudence dictates that all chemical substances, both old and new, be subjected to at least basic screening for neurotoxicity in the light of expected use and exposure. However, the sheer number of untested chemicals constitutes a practical problem of daunting magnitude for neurotoxicology. Given the number of untested chemicals and current limitations on resources, they cannot all be tested for neurotoxicity in the near future. Testing procedures designed for neurotoxicologic evaluation that have been developed so far might be reasonably effective, but are so resource-intensive that they could not be applied to all untested chemicals.

A rational approach to neurotoxicity testing must contain the following elements:

• Sensitive, replicable, and cost-effective neurotoxicity tests with explicit guidelines for evaluating and interpreting their results.
• A logical and efficient combination of tests for screening and confirmation.
• Procedures for validating a neurotoxicologic screen and for guiding appropriate confirmatory tests.
• A system for setting priorities for testing.

This chapter discusses systematic assessment of chemicals for neurotoxic hazards. It begins by describing biologic and public-health issues that are peculiar to neurotoxicology. It then presents a review of current techniques for neurotoxicologic assessment to address the question: "How effective are current test procedures for identifying neurotoxic hazards?" Concomitantly, the review seeks to identify gaps and unmet needs in current neurotoxicity test procedures. Finally, the chapter describes how the regulatory agencies currently evaluate potential neurotoxicity and how the tests reviewed could be combined in an improved neurotoxicity testing strategy.

A byproduct of the testing strategy outlined in this chapter will be the development of an array of biologic markers of neurotoxicity. These markers can be used in future studies of experimental animals, as well as in clinical and epidemiologic studies of humans exposed to neurotoxicants, as was proposed in Chapter 3.

APPROACH TO NEUROTOXICITY TESTING

Difficulties in Neurotoxicity Testing

Neurotoxicity testing is relatively new. Although its rapid development is noteworthy, its progress has been constrained by several factors that complicate neurotoxicologic assessment. Some of the complexities, such as sex- or age-related variability in response, are common to all branches of toxicology. Neurotoxicology, however, faces unique difficulties, because of several characteristics that make the nervous system particularly vulnerable to chemically induced damage (Chapter 2). Those characteristics include the limited ability of the nervous system to repair damage, because of the absence of neurogenesis in adults; the precarious dependence of axons and synaptic boutons on long-distance intracellular transport; the system's distinct metabolic requirements; the system's highly specialized cellular subsystems; the use of large numbers of chemical messengers for interneuronal communication; and the complexity of the nervous system's structural and functional integration. The nervous system exhibits a greater degree of cellular, structural, and chemical heterogeneity than other organ systems. Toxic chemicals potentially can affect any functional or structural component of the nervous system—they can affect sensory and motor functions, disrupt memory processes, and cause behavioral and neurologic abnormalities. The large number of unique functional subsystems suggests that a great diversity in test methods is needed to ensure assessment of the broad range of functions susceptible to toxic impairment. The special vulnerability of the nervous system during its long period of development is also a critical issue for neurotoxicology.

Despite the inherent difficulties of neurotoxicity testing, some validated tests have been developed and implemented. Testing strategies must take those facets of the nervous system into account, and they must consider a number of variables known to modify responses to neurotoxic agents, such as the developmental stage at which exposure occurs and the age at which the response is evaluated. The issue of timing is complex. During brain development, limited damage to cell function—even reversible inhibition of transmitter synthesis, for instance—can have serious, long-lasting effects, because of the trophic functions of neurotransmitters during neuronal development and synaptogenesis (Rodier, 1986). Kellogg

(1985) and others have shown the striking differences between the effects of perinatal exposure of animals to some neuroactive drugs (e.g., diazepam) and the effects of exposure of adult animals to the same drugs. Other agents (e.g., methylmercury and lead) are toxic at every age, but are toxic at lower doses in developing organisms. In addition, in developing animals, the blood-brain barrier might not be sufficiently developed to exclude toxicants. Stresses later in development might lead to the expression of relatively sensitive effects that were latent or unchallenged at earlier stages of development. During senescence, the CNS undergoes further change, including a loss of nerve cells in some regions. CNS function in senescence could be vulnerable to cytotoxic agents that, if encountered earlier in development, might have been protected against by redundant networks or compensated for by "rewiring" of networks (Edelman, 1987). To address those complex issues, testing paradigms that incorporate both exposures and observations during development and during aging need to be considered. The disorders might be acute and reversible or might lead to progressive disorders over the course of chronic exposure. More sensitive biologic markers as early indicators of neurotoxicity are urgently needed. Opportunities should be exploited for detecting neurotoxicity when chronic lifetime bioassays are conducted for general toxicity or carcinogenicity.

In the design of neurotoxicity screens, no test can be used to examine all aspects of the nervous system. The occurrence of an effect of a chemical on one function of the nervous system will not necessarily predict an effect on another function. Therefore, it is important to use a variety of initial tests that measure different chemical, structural, and functional changes to maximize the probability of detecting neurotoxicity or to use tests that sample many functions in an integrated fashion.

The Testing Strategy

Efficient identification of potential hazards warrants a tiered testing strategy. The first tier of testing (the screen) need not necessarily be specifically predictive of the neurotoxicity likely to occur in humans, unless regulatory agencies are to use the results for direct risk-management decisions. The tests in later tiers are essential to assess specificity and confirm screening results and are appropriate for defining dose-response relationships and mechanism of action (Tilson, 1990a). Screening tests would be followed, as appropriate, by more specific assessments of particular functions. Such an approach permits a decision to be made about whether to continue testing at each step of the progression. In the case of chemicals already in use (in which case people are already being exposed and financial consequences might be considerable), detailed testing to determine mechanisms of toxicity would be pursued when screening tests revealed neurotoxic effects; positive findings on a chemical undergoing commercial development might trigger its abandonment with no further testing.

Testing at the first tier is intended to determine whether a chemical has the potential to produce any neurotoxic effects—i.e., to permit hazard identification, the first step of the NRC (1983) risk-assessment paradigm. The next tier is concerned with characterization of neurotoxicity, such as the type of structural or functional damage produced and the degree and location of neuronal loss. During hazard characterization (the second step of the NRC paradigm), tests are used to study quantitative relationships between exposure (applied dose) and the dose at the target site of toxic action (delivered dose) and between dose and biologic response. The third and final tier of neurotoxicity testing is the study of mechanism(s) of action of chemical agents.

The decision to characterize a chemical through second-tier testing might be moti-vated by structure-activity relationships, existing data that suggest a chemical is neurotoxic, or reports of neurotoxic effects in humans exposed to the chemical, in addition to the results of first-tier testing. Testing at this second tier can help to resolve several issues, including whether the nervous system is the primary target for the chemical and what the dose-effect and time-effect relationships are for relatively sensitive end points. Such tests can also be useful in the determination of a no-observed-adverse-effect level (NOAEL) or lowest-observed-adverse-effect level (LOAEL). Experiments done at the third tier can also examine the mechanism of action associated with a neurotoxic agent; they will often involve neurobehavioral, neurochemical, neurophysiologic, or neuropathologic measures. They might also suggest biologic markers of neurotoxicity for validation and use in new toxicity tests and in epidemiologic studies.

Characteristics of Tests
Useful for Screening

Like any other toxicity test, screening tests for neurotoxicity should be sensitive, specific, and valid. Sensitivity is a test's ability to detect an effect when it is produced (the ability to register early or subtle effects is especially desirable). Sensitivity depends on inherent properties of the test and on study-design factors, such as the numbers of animals studied and the amount and duration of exposure. Specificity is a test's ability to respond positively only when the toxic end point of interest is present. Specificity and sensitivity are aspects of accuracy. An inaccurate test fails to identify the hazardous potential of some substances and incorrectly identifies as hazardous other substances that are not. In statistical terms, the failure to identify a hazard is a false-negative result, and the mistaken classification of a safe substance as hazardous is a false-positive result. Increasing test specificity reduces the incidence of false positives, but often has the unwanted consequence of increasing the incidence of false negatives (decreasing sensitivity); increasing test sensitivity reduces the incidence of false negatives, but often increases the incidence of false positives.

An ideal screen would have broad specificity, so that it could detect all aspects of nervous system dysfunction. In practice, no screening procedure is likely to provide the desired coverage without producing false positives. The sensitivity of the first tier might be maximized by a battery of screening tests that are individually quite specific. Several durations of exposure, long postexposure observation periods, and lifetime tests might all be necessary to cover the possible manifestations of neurotoxicity. In a tiered test system, very high sensitivity of the screening tier is usually considered essential; the loss of specificity must be compensated for in later tiers, to reveal the false positives. If, however, product development will be aborted on any indication of neurotoxicity, false positives could have high social costs. In the later tiers, narrower specificity is appropriate to characterize a suspected toxicant or to establish its mechanism of action.

Validation is the process by which the credibility of a test is established for a specific purpose (Frazier, 1990). It entails demonstrating the reliability of the test's performance in giving reproducible results within a laboratory and in different laboratories and giving appropriate results for a control panel of substances of known toxicity. The usefulness of the results of a neurotoxicity screening test depends on a positive outcome's being strongly correlated with a neurotoxic effect that is actually caused by exposure to the test substance. Direct mechanistic causality is not essential for their interpretation, although direct insight into mechanism would be valuable, as is sought when developing a biologic marker of effect.

Validation of a test system should also include demonstration that positive test results indicate that neurotoxic effects would occur if humans were exposed to the substance. Predicting an influence on human affect or cognition with nonhuman test systems is challenging, but possible. Many animal testing models do produce effects that correspond to those seen in humans exposed to the same substance, but a result would also be considered valid if it correctly predicted any neurotoxic effect in the human population. For example, consider a screening test for an effect produced only at high doses that is consistently correlated with a milder, presumably precursor effect that occurs at lower doses. The easily detectable effect occurring at high doses in experimental animals might never be observed in humans, whose exposure would never be extreme, but its occurrence in the animal model might indicate that low-level exposure in humans could produce a more subtle toxic effect. O'Donoghue (1986) has noted that chemicals that damage axons are often metabolic poisons that produce a retardation of weight gain without decreasing food intake—a more easily observed end point. It is important to recognize that chemicals can adversely affect multiple organ systems, and an effect in other organs might influence some measures of neurotoxicity. In vitro assay systems often measure end points that appear unrelated to neurologic functioning in whole animals, but detect alterations to crucial underlying mechanistic processes.

Application of the Screen and Later Tiers

Given the enormous number of substances that have not been tested for neurotoxicity (or any form of toxicity), some characteristics of chemicals, such as their structure or production volume, must contribute to the determination of priority for screening. Confidence in the combined sensitivity of tests composing a screening battery should be high enough for negative findings to be reliably regarded as acceptable evidence that a substance is unlikely to have neurotoxic activity without the need for additional testing. If a new chemical gives positive results, there might nevertheless be academic or commercial interest in pursuing a risk assessment, or its development for use might be abandoned without further testing. Detailed testing to characterize neurotoxicity revealed by screening procedures will be more common for existing substances with commercial value or wide exposure. In either case, the path followed through the tiers of testing would be contingent on a chemical's unfolding toxicity profile, including types of toxicity other than neurotoxicity. A feasible screening system must strike a balance among the amount of time and expense that society is willing to expend in testing, its desire for certainty that neurotoxic substances are being kept out of or removed from the environment, and its interest in gaining benefits from various types of chemicals.

CURRENT METHODS BASED ON STRUCTURE-ACTIVITY RELATIONSHIPS

Given the overwhelming lack of epidemiologic or toxicologic data on most chemical substances and the need to develop rational strategies to prevent adverse health effects of both new and existing chemicals, attempts have been made in many subfields of toxicology to generate predictive strategies based primarily on chemical structure. The basis for inference from structure-activity relationships (SARs) can be either comparison with structures known to have biologic activity or knowledge of structural requirements of a receptor or macromolecular site of action. However, given the complexity of the nervous system and the lack of information on biologic mechanisms of neurotoxic action, SARs should be regarded as, at best, pro-

viding information that might be useful to identify potentially neuroactive substances. SARs are clearly insufficient to rule out all neurotoxic activity; it is not prudent to use them as a basis for excluding potential neurotoxicity. Caution is warranted in interpreting SARs in anything other than the most preliminary analyses. An intelligent use of SARs requires detailed knowledge not only of structure, but also of each critical step in the pathogenetic sequence of neurotoxic injury. Such knowledge is still generally unavailable. SAR evaluations form a major basis for EPA and FDA decisions on whether to pursue full neurotoxicity assessments, so it must be concluded that the approach of these agencies cannot be expected to provide an adequate screen for identifying neurotoxicants.

SAR approaches are more successful when the range of possible sites of action or mechanisms of action is narrow. Thus, SARs have had more use in relation to carcinogenicity and mutagenicity than in other kinds of toxicity. The SAR approaches used in the development of novel neuropharmacologic structures deserve consideration in neurotoxicology, but they must rest on fuller understanding of neurotoxic mechanisms.

SARs have not been used extensively in neurotoxicology, for several reasons. Many agents are neurotoxic—from elements, such as lead and mercury, to complex molecules, such as MPTP and neuroactive drugs. Neurotoxicants have so many potential targets that it is difficult to rule out any chemical a priori as not likely to be neurotoxic. Finally, for most neurotoxicants, even those which are well characterized, data on the mechanism of action at the target site are insufficient for the elucidation of useful SARs.

SARs have proved useful in some cases—usually, within particular classes of compounds on which some mechanistic data are available. Extensive studies of anticholinesterase organophosphates and carbamates, for example, have led to a mechanistic model that requires a structure that binds to a specific site on the cholinesterase molecule and to the establishment of an SAR for cholinesterase inhibition. The SAR is relatively straightforward, in that enzyme inhibition is the primary action that leads to various symptoms of poisoning. Another series of studies has been conducted in an attempt to develop an SAR for the pyrethroid insecticides. Although the nerve membrane sodium channel has been identified as the critical target site of pyrethroids, studies have failed to establish a clear-cut SAR, despite the large number of pyrethroids synthesized and tested. However, the importance of the d-cyano group in distinguishing two behavioral manifestations (choreoathetosis and salivation versus tremor) and the importance of the ester moiety in pyrethroid SAR are well documented (Aldridge, 1990). A more comprehensive SAR for pyrethroids is lacking partly because most studies were performed with insecticidal action as a measure of activity, not the interaction with the sodium channel. Action at the target site must be known to establish an SAR. However, insecticidal potency involves factors other than recognition at the target site, including absorption and metabolism. As the mechanism of action of n-hexane and methyl n-butylketone neurotoxicity becomes more clear, so does the capacity to look at an alkane or ketone and predict its neurotoxic potential. The critical issue in the neurotoxicity of such a chemical is whether it will be metabolized to a γ-diketone; α-, β-, and δ-diketones are less likely to be neurotoxic (Krasavage et al., 1980; Spencer et al., 1978), but most γ-diketones and γ-diketone precursors cause a specific type of neurotoxicity, a motor-sensory peripheral neuropathy caused by neurofilament abnormalities. The gamma spacing of the two ketone groups allows formation of the five-membered heterocyclic ring, the pyrrole, except where steric hindrance is present (Genter et al., 1987).

CURRENT IN VITRO PROCEDURES

In vitro procedures for testing have practical advantages, but studies must be done to correlate their results with responses in whole animals. One advantage of validated in vitro tests is that they minimize the use of live animals. Some of the more developed in vitro tests might be simple and might not have to be conducted by highly trained personnel, but, as with many in vivo tests, the analysis and interpretation of results is likely to require expertise. Some in vitro tests lend themselves to computer-controlled automated operation, as do some well-developed, highly sophisticated behavioral tests (see, for example, Evans, 1989), and that results in savings in time and expense and allows testing of large numbers of substances. Experience with the Ames test for mutagenesis confirms the advantages of in vitro procedures, but also illustrates the problems that arise when an assay is used to predict an end point that is not exactly what it measures (e.g., carcinogenicity, rather than specific sorts of genotoxicity).

Biochemical Assays

Biochemical assays epitomize the advantages of in vitro tests. Their current usefulness, however, is limited to substances in a very few chemical classes. Their usefulness will no doubt increase as the molecular basis of action of other classes of neurotoxicants becomes known.

Neurotoxic chemicals exert their effects via specific molecular interactions with biologic targets. For a few toxicants, the molecule that is affected is well established, and it is possible to investigate the potential toxicity of other substances in the same class with a biochemical assay. An example of this approach is the enzyme-inhibition assay of organophosphorus (OP) esters. Some OP esters cause a distal polyneuropathy that is not evident until several days after exposure (OP-induced delayed neuropathy, or OPIDN). There is now good evidence (Abou-Donia, 1981) that the capacity of a given OP ester to cause OPIDN correlates strongly with the relative inhibition of a CNS enzyme activity called neurotoxic esterase. The enzyme activity is easily measured in a test-tube assay, and the addition of OP esters and OP-like compounds to the assay allows one to screen for ability to inhibit the enzyme and thus for its potential to produce OPIDN (Johnson, 1977). The ease of this assay is complicated, however, by the precautions necessary to protect the technicians performing the assay from exposure to potentially highly neurotoxic substances. Many OP compounds also produce an immediate toxic effect via inhibition of another enzyme, acetylcholinesterase. A test-tube assay of that enzyme's activity can serve as a screen for such acute neurotoxicity.

Tissue Culture

The study of in vitro systems has provided much fundamental information of value in understanding the nervous system in vivo, and in vitro investigations have sometimes been invaluable in guiding in vivo neurotoxicologic research. The physiologically excitatory actions of some amino acids (for example, glutamate and aspartate) on neurons can become pathologic, and brain lesions can be produced by amino acid analogues, such as kainate or quisqualate. In vitro systems are being used extensively to study the mechanisms of the neuronal injury produced by those agents and to identify antagonists that might be useful in reducing excitotoxic brain damage. In vitro studies can also identify compounds with a high neurotoxic potential that might merit study in intact organisms. An example is β-N-methylamino-L-alanine (BMAA), a constituent of a dietary plant, the false sago palm. BMAA was initially identified as an ex-

citotoxic amino acid in cultures of tissue from spinal cord and cerebral cortex. That led to experiments with primates that showed that BMAA produces motoneuronal lesions in the cortex and spinal cord.

A broad range of tissue-culture systems are available for assessing the neurologic impact of environmental agents. Although those systems are not now used for hazard detection, they can be used to characterize chemical-induced effects. They can be classified according to their increasing complexity, from cell lines to organ cultures.

• *Cell lines.* Neuronal and glial cell lines have been used in many neurobiologic studies and are valuable in neurotoxicology. They consist of populations of continuously dividing cells that, when treated appropriately, stop dividing and exhibit differentiated neuronal or glial properties. Various neuronal lines, for instance, develop electric excitability, chemosensitivity, axon formation, transmitter synthesis and secretion, and synapse formation. Large quantities of cells can be generated routinely to develop extensive dose-response or other quantitative data. For example, the dose range over which a group of chemicals affect cell differentiation and proliferation was established in neuroblastoma cells (Stark et al., 1989). Those are tumor cells, however, so the interpretation of such data with regard to toxicity in the intact nervous system must be guarded. A culture system for nontransformed neural cells was recently announced (Ronnett et al., 1990).

• *Dissociated cell cultures.* When neural tissue, typically from fetal animals, is dissociated into a suspension of single cells, and the suspension is inoculated into tissue-culture dishes, the neurons and glia survive, grow, and establish functional neuronal networks. Such preparations have been made from most regions of the CNS and exhibit highly differentiated, site-specific properties that constitute an in vitro model of different portions of the CNS. Most of the neuronal transmitter and receptor phenotypes can be demonstrated, and a variety of synaptic interactions can be studied. Glial cells are also present, and neuroglial interactions are a prominent feature of the cultures. A substantial battery of assays (neurochemical and neurophysiologic) is now available to assess the development of the cultures and to indicate toxic effects of test agents added to the culture medium. Relatively pure populations of different cell types can be isolated and cultured, so that effects on specific cell types can be assessed independently. Pure glial cells or neurons, or even specific neural categories, can be prepared in this way and studied separately, or interaction between neurons and glial cells can be studied at high resolution. The neurobiologic measures used to assess the effect of any agent can be very specific (for example, activity of a neurotransmitter-related enzyme or binding of a receptor ligand) or global (for example, neuron survival or concentration of glial fibrillary acidic protein). The two-dimensional character of the preparations makes them particularly suited for morphologic evaluation, and detailed electrophysiologic studies are readily performed. The toxic effects and mechanisms of anticonvulsants, excitatory amino acids, and various metals and divalent cations have been assessed with these preparations. The cerebellar granule cell culture system, for example, has been exploited recently in studies of the mechanism of alkyllead toxicity (Verity et al., 1990).

• *Reaggregate cultures.* A related but distinct preparation made from single-cell suspensions of neural tissue is the reaggregate culture. Instead of being placed in culture dishes and allowed to settle onto the surface of the dishes, the cells are kept in suspension by agitation; under appropriate conditions, they stick to one another and form aggregates of controllable size and composition. Typically, the cells in an aggregate organize themselves and exhibit intercellular relations that are a function of and bear some resemblance to the brain region that was the source of the cells. The cells establish a three-dimensional, often laminated structure, perhaps approximating the in vivo nervous system more closely than do the dissociated cultures grown on the surface of a dish. Reaggregate cultures lend themselves to large-scale, quantitative experiments in which neurobiologic variables can be examined, although morphologic and ligand-binding studies are performed less readily than with surface cultures.

• *Explant cultures.* Organotypic explant cultures are even more closely related to the intact nervous system. Small pieces or slices of neural tissue are placed in culture and can be maintained for long periods with substantial maintenance of structural and cell-cell relations of intact tissue. Specific synaptic relations develop and can be maintained and evaluated, both morphologically and electrophysiologically. Because all regions of the nervous system are amenable to this sort of preparation, it is possible to analyze toxic agents that are active only in specific regions of the central or peripheral nervous system. Explants can be made from relatively thin slices of neural tissue, so detailed morphologic and intracellular electrophysiologic studies are possible. Their anatomic integrity is such that they capture many of the cell-cell interactions characteristic of the intact nervous system while allowing a direct, continuing evaluation of the effects of a potentially neurotoxic compound added to the culture medium. The process of myelination has been studied extensively in explant cultures, and considerable neurotoxicologic information has been gained. As noted above, the pathogenetic actions of excitatory amino acids normally active in the nervous system, as well as such analogues as the neurotoxin BMAA, have been revealed by experiments with organotypic cultures.

• *Organ Cultures.* A preparation similar to an explant culture is the organ culture, in which an entire organ, such as the inner ear or a ganglion, rather than slices or fragments, is grown in vitro. Obviously, only structures so small that their viability is not compromised can be treated in this way.

The advantages of the various types of in vitro systems are summarized in Table 4-1. Most in vitro preparations are made from young, usually prenatal animals. (But cultures derived from human neural tissue have been the object of a number of studies.) Typically, a period of rapid change and development occurs immediately after the cultures are initiated, and conditions become much more stable if the cultures are maintained for weeks or months. Thus, the preparations can be used to study neurotoxic effects that might be specific to developing nervous tissue and to compare the effects of agents in developing stable tissue.

In general, the technical ease of maintaining a culture varies inversely with the degree to which it captures a spectrum of in vivo characteristics of nervous system behavior. The problem of biotransformation of potentially neurotoxic compounds is shared by all, although the more complete systems (explant or organ cultures) might alleviate this problem in specific instances. In many culture systems, complex and ill-defined additives—such as fetal calf serum, horse

serum, and human placental serum—are used to promote cell survival. A number of thoroughly described synthetic media are now available, however, and such fully defined culture systems can be used where necessary. Indicators of neuronal and glial function, and hence indicators of neurotoxicity, are outlined in Table 4-2.

Plausibility of an In Vitro Screening Battery

A broad range of in vitro systems are now available for studying development of the nervous system and the normal function of neurons and glial cells. The possible neurotoxic impact of any chemical on any specific neurobiologic variable could, in principle, be screened with an appropriate set of in vitro tests. In practice, of course, because the number of potential neurobiologic end points to be measured is so large, screening for the effects of any agent on all of them would be prohibitively expensive in time and money. The question, then, is whether a feasible battery of tests will pick up an acceptably large percentage of toxic chemicals (while generating an acceptably low percentage of false positives). Ideally, the screen would

TABLE 4-1   In Vitro Neurobiologic Test Systems

Culture Type	Comparablity to In Vivo System[a]	Mechanistic Analysis[a]	Features
Cell line	+	+ + + +	Large quantities of material
Dissociate cell culture	+ +	+ + +	Good accessibility for study
Reaggregate culture	+ + +	+ +	Three-dimensional structure
Explant culture	+ + + +	+ +	Good approximation of intact sytems
Organ culture	+ + + +	+	Less long-term survival than other models

[a]Number of pluses reflects estimated relative advantage and represents the committee's judgment.

TABLE 4-2   Markers for Assessing Neurotoxicity in In Vitro Systems

	Degree of Difficulty*
General neuronal measures	
Cell number	1
Tetanus-toxin binding	2
Neurofilament protein	3
Neuronal structure	2
General glial measures	
Glial fibrillary acidic protein	2
Oligodendrocyte probe	3
Transmitter systems	
Amino Acid	
Excitatory	3
Inhibitory	1
Cholinergic	
Choline acetyltransferase	1
Muscarinic and nicotinic receptors	3
Aminergic	
Norepinephrine	2
Serotonin	2
Dopamine	2
Peptidergic	
Vasoactive intestinal peptide	3
Substance P	3
Enkephalin	3
Cell biologic responses	
Second messengers	
Cyclic nucleotide	3
Phosphoinositide turnover	3
Phosphorylation	3
Calcium-dependent transmitter release	3
Voltage-dependent NA^+ or Ca^{2+} uptake	2

*The higher the number, the more difficult to measure. This represents the committees judgment.

provide direction for the more intensive study of substances that it identifies as having neurotoxic potential.

To what extent could an in vitro system provide such a screening instrument? Two basic questions are associated with the use of in vitro tests for that purpose:

· What indicators of neuronal (or glial) damage would be sufficiently general to be useful?

· What specific test systems or combinations would be adequate to cover a number of different and differentially site-specific neurotoxic agents?

The first question might be answered by a combination of assays that would include general indicators of neuronal and glial survival and a few more specific indicators. Counts of numbers of surviving neurons (or glia) and biochemical measurements of tetanus-toxin-specific or sodium-channel-specific ligand binding could be used as general indicators. Uptake of γ-aminobutyric acid (GABA), benzodiazepine binding, and cholinergic and aminergic markers could be used, depending on the neural system chosen, to get some indication of neurotransmitter-related functions. If a chemical has a single very specific neurologic target, this would in general be missed, but it might be anticipated that such a specific effect would be accompanied by more general secondary neuropathologic consequences. For instance, if spinal-cord or cerebral cortical cell cultures are exposed to the specific voltage-dependent sodium channel blocker tetrodotoxin for 4-5 days, the decrease in electric activity kills about half the neurons.

As to the second question, the choice of culture systems to be used is difficult. It is axiomatic that no cell culture represents a normal nervous system. Even in cocultured explants, the normal connections among cells are disrupted. Specific neurobiologic properties have been shown *not* to be expressed in various in vitro preparations where they might be expected. For instance, in dissociated hippocampal pyramidal cells, serotonergic responses cannot be demonstrated; the responses appear during development in vivo, but appropriate signals that induce their expression are evidently lacking in vitro. It is impossible to predict how such departures from normality will influence the screen's ability to detect the effects of a test substance. Some compromise between comprehensiveness and fiscal feasibility would have to be made. Neural and glial cell lines are available and relatively straightforward technically. Robust versions of cerebral-cortex, spinal-cord, and subcortical systems are available either as dissociated preparations or in explant form. Indeed, essentially all regions of the nervous system are being grown in vitro in one form or another. If a candidate neurotoxic material has properties that suggest site-specific activity, obviously one would include its putative target in a test battery. If no such information is available, then some arbitrary panel of test systems would have to be used.

Even if these two questions are answered, cell-culture techniques have several remaining disadvantages when used as screens. A given toxicant may require metabolism outside the nervous system to produce a toxic metabolite, so exposure to the toxicant in vitro may give a false-negative result; conversely, the chemical might be detoxified before reaching the nervous system. A related disadvantage might be low solubility of a given toxicant in an aqueous culture medium, which could limit the quantity of toxicant to which the cells are actually exposed. A related problem concerns the lack of a blood-brain barrier in in vitro experiments. Toxicity could be attributed to a compound that would not reach the brain, either at all or in sufficient concentration to cause toxicity, in in vivo exposure. Last, and most important, the more complex functions of the nervous system are properties of assemblies of neurons. Learning, memory, emotions, the coordination of movement,

and homeostatic regulation, for instance, cannot be studied in vitro.

Despite those disadvantages, some applications of cell cultures are nearly ready for use in neurotoxicity screening. Several indicators of glial and neuronal function in vitro can be used as biologic markers of effect (Table 4-2). In vitro systems have been extremely useful in identifying and analyzing excitotoxic materials. Also, in vitro demonstrations of neurotoxicity of anticonvulsants based on both general and specific indicators have been corroborated by clinical studies of IQ decrements related to phenobarbital treatment in children. Many neurotoxicants attack the glial cells that form myelin sheaths around axons, and cell cultures can demonstrate chemically induced myelin abnormalities; this has been shown with the myelin abnormalities produced by methylmercury (Kim, 1971), thallium (Spencer et al., 1985), triethyltin (Graham et al., 1975), and several other substances. Tests on cultures can also detect toxic damage to neurons themselves. For example, aluminum neuronopathy is reproduced in cultures of dorsal root ganglion cells (Seil et al., 1969), and the axonal degeneration characteristic of γ-diketone toxicity is reproduced in organotypic cocultures (Veronesi et al., 1978). Furthermore, as noted in Chapter 2, the cells of the CNS are uniquely dependent on a high rate of glucose metabolism; many neurotoxicants impair neural glucose metabolism (Damstra and Bondy, 1980), and this toxic mechanism could be reflected in a decrease in the viability of cultured cells.

A somewhat arbitrary, but specific, protocol for developing a set of in vitro tests for screening is presented, for illustrative purposes, in Table 4-3. It is an empirical question whether such a battery of in vitro tests could contribute importantly to neurotoxicologic screening of a broad range of environmental agents. Research needs to be done with known neurotoxic agents and related compounds to see whether culture systems can reliably identify dangerous

TABLE 4-3   Proposed Protocol for Developing an In Vitro Neurotoxicity Screening System

1. Test candidates representing various types of potential neurotoxicants, e.g., anticholinesterases, excitatory amino acids, trimethyl tin.

2. Use three concentrations, e.g., less than, equal to, and greater than in vivo toxic concentrations.

3. Use three types of preparation, e.g., PC12 cells, dissociated neurons and glia of cerebral cortex, and explants of dorsal root ganglia and spinal cord.

4. Study developing (0-2 weeks in vitro) and mature (> 2 weeks in vitro) preparations.

5. Evaluate appropriate end points, such as cell counts, glial fibrillary acid protein, structure, choline acetyltransferase, and glutamic acid decarboxylase.

6. Use inactive isomers or analogues as negative control agents.

substances and distinguish them from related, but inactive or less active, substances. Analyses of the contributions of short-term in vitro tests to neurotoxicity are for less developed and for fewer in number than analyses of in vitro tests in other fields, e.g., carcinogenesis (Lave and Omenn, 1986).

It should be possible to define the goals of such research to be finite and realizable. The question may be framed as follows: What is the smallest set of tests that gives positive indicators of neurotoxicity for all of a control panel of compounds known to have diverse neurotoxic properties? Of equal importance, and perhaps more difficult, would be the evaluation of presumably nonneurotoxic substances to preclude a high rate of false positives. One might start with a fairly inclusive set of test cultures, such as of cerebral cortex, brain stem, cerebellum, spinal cord, sympathetic ganglion neurons, and a continuous cell line, such as PC12 cells. Global assays, such as cell counts and tetanus-toxin binding, would be initial indicators, and specific neurotransmitter-related probes could be used as appropriate. The essential feature of such an effort is the reliability of the culture systems. Reliability typically is achieved only after a fairly long period of use in a given laboratory.

It will be difficult to cover all end points of concern with a modest number of in vitro assays. Whole-animal tests have the potential of exploiting the integrative nature of behavior, thereby covering a diversity of adverse end points that might result from exposure to neurotoxic agents. It might be more efficient to screen with in vivo tests and use an in vitro approach when concentrating on specific mechanisms of action. A goal of research directed at development of adequate neurotoxicologic methods should be to combine behavioral testing on intact animals with selected in vitro tests. Can some behavioral tests be replaced by less expensive in vitro tests without loss of diagnostic power? Can individual behavioral tests function as well as or better than groups of in vitro tests? One might start with the study of Tilson (1990a) as a base and complement it with in vitro study of the same chemicals. The results might lead to a choice of tests that will yield an optimal combination of broad coverage and low cost.

CURRENT IN VIVO PROCEDURES

Neurotoxic effects on complex integrative functions—such as motor performance, sensory acuity, memory, and cognitive processes—can be detected only in vivo. Moreover, the neural activities that mediate integrative function involve large numbers of neurons relatively distant from each other. Integration may be disrupted by the removal and isolation of neural tissue that is necessary for some in vitro techniques, but neurotoxic effects of relatively low exposures involving integrative functions can be detected in whole animals.

Species differ in their susceptibility to various toxic agents and the degree to which their nervous systems resemble that of humans. Tests using species closest to humans would logically yield the data about which we could feel most confident; however, various considerations—including cost, ethics, and the extent of pre-existing data bases—favor the use of small laboratory rodents, such as mice and rats, for in vivo hazard identification and characterization. Table 4-4 lists commonly reported neurotoxic effects of several classes of toxic chemicals in humans and animals. It is apparent from the list that motor disturbances, mood alteration, and sensory abnormalities are especially common in humans, but that the findings in animals are only sometimes comparable. Species differences might account for some of these discrepancies; however, it is not clear that animal tests always are intended to be or can be exactly comparable with human tests.

Testing for neurotoxicity in humans implies that exposure has occurred. Such testing is therefore considered a means of secondary prevention, so the testing methods specifically for humans are discussed in Chapter 5.

Behavioral Assessment

Chemical-induced functional alterations of the nervous system are often assessed with behavioral techniques. Perhaps the greatest scientific challenge to neurotoxicology is to integrate observations of behavior with other aspects of neurobiology—such as morphology, neurochemistry, and neurophysiology—to develop a unified theory not only of toxicity, but also of the nervous system. By itself, behavior is an important end point, even if its biologic substrates have not been clearly identified. Numerous behavioral techniques are available to measure chemical-induced alterations in sensory, motor, autonomic, and cognitive function (Table 4-5).

Behavioral methods differ greatly in their complexity and specificity. At one extreme, some methods (e.g., some observational tests) can be applied broadly and routinely to assess the neurotoxicity of a wide array of chemicals and chemical exposures; that is, they are useful in hazard identification. Such tests typically incorporate responses already well established in an organism. Other methods require instrumentation or training of the animals before chemical exposure; although they might not be appropriate for the routine screening (tier 1) of chemicals for neurotoxicity, they could be used for characterization of toxicant-induced effects (tier 2).

Functional Observational Batteries

Functional observational batteries (FOBs) are designed to detect and measure major overt neurotoxic effects. Several have been used, each consisting of tests generally intended to evaluate various aspects of sensorimotor function (EPA, 1985; Haggerty, 1989; Kulig, 1989; Moser, 1989; O'Donoghue, 1989). FOB tests are essentially clinical examinations that detect the presence or absence, and in some cases the relative degree, of specific neurologic signs.

Screening neuroactive chemicals with an FOB is well established. Irwin (1968), for example, described a series of tests for evaluating the effects of drugs in mice and showed how different drugs produced different patterns of effects that could be easily recognized. Gad (1982) described a battery

TABLE 4-4 Neurotoxic Effects of Representative Agents in Humans and Animals

Chemical Class	Representative Agents	Neurotoxic Effects in Humans	Neurotoxic Effects in Animals
Solvents	Hexane, acrylamide	Ataxia, tremor, paresthesia, hypersomnia, slurring of speech, delirium and hallucinations	Loss of fine motor control, weakness, sensory system disturbance
	Carbon disulfide	Anosmia, paresthesia, depression, anxiety, psychoses	Sensory system disturbance
Organochlorine insecticides	Chlordecone, DDT	Ataxia, tremor, slurring of speech, euphoria and excitement, nervousness and irritability, depression and anxiety, mental confusion, memory disorders	Tremor, hyperreflexia, impaired acquisition
Organophosphate esters	Parathion, paraoxon	Ataxia, paresthesia, insomnia, slurring of speech, tinnitus, amblyopia, nystagmus, abnormal pupil reactions, nervousness and irritability, depression and anxiety, psychoses, memory disorders	Weakness, sensory system disturbance, autonomic dysfunction, impaired acquisition
Organometals	Methylmercury	Ataxia, myoclonus, paresthesia, insomnia, slurring of speech, hearing loss, abnormal pupil reactions, mental confusion	Weakness, sensory system disturbance, visual deficits, learning deficits

TABLE 4-4 (Continued)

Chemical Class	Representative Agents	Neurotoxic Effects in Humans	Neurotoxic Effects in Animals
Heavy metals	Inorganic lead	Ataxia, tremor, pathologic reflexes, paresthesia, hearing loss, abnormal pupil reactions, depression and anxiety	Motor disturbances, sensory system disturbance
	Mercury vapor	Facial tic, tremor, insomnia, amblyopia, depression and anxiety	Motor disturbances, tremor
	Manganese	Tremor, paresthesia, hypersomnia, euphoria and excitement, delirium and hallucinations, memory disorders	Tremor, motor disturbances
	Cadmium	Anosmia	Anosmia
	Arsenic	Hyperesthesia	Hyperesthesia

Source: Modified from Tilson and Mitchell (1984).

TABLE 4-5 Examples of Behavioral Measures of Functional Neurotoxicity

Function Affected	Signs and Symptoms	Animal Test[a]	Tier of Testing	
			1: Hazard Identification	2: Characterization
Sensory	Abnormalities of smell, vision, taste, hearing	FOB	X	
		Reflex modification		X
		Conditioned discrimination		X
Motor	Muscle weakness	FOB	X	
		Grip strength	X	
		Hindlimb splay	X	
		Motor discrimination		X
		Swimming endurance	X	
		Suspension from bar	X	
	Tremor	FOB	X	
		Spectral analysis		X
	Convulsions	FOB	X	
	Incoordination	FOB	X	
		Negative geotaxis	X	
		Rotorod	X	
		Inclined screen	X	
		Motor discrimination		X
	Hypoactivity or hyperactivity	Motor activity	X	

TABLE 4-5 (Continued)

Function Affected	Signs and Symptoms	Animal Test[a]	Tier of Testing	
			1: Hazard Identification	2: Characterization
Autonomic	Abnormalities of sweating, temperature control, gastrointestinal function	FOB	X	
Cognitive	Disruption of learned behavior	Schedule-controlled operant behavior (SCOB)		X
	Learning and memory	Habituation	X	
		Classical		X
		Instrumental		X

[a]FOB = functional observational battery

of tests for assessing the neuromuscular effects of industrial chemicals in rats. More recently, Moser and colleagues (Moser et al., 1988; Moser, 1989) developed a similar battery for assessing the neurobehavioral effects of a broad range of industrial and pesticidal chemicals in rats.

FOBs are sets of observations and tests each made on individual experimental animals; e.g., the FOB suggested by Moser (1989) is presented in Table 4-6. It is assumed that many of the individual observational components overlap in the neurologic functions that they assess (autonomic function, motor function, equilibrium, excitability, and sensorimotor reflexes). Therefore, if several unrelated observed end points in an entire FOB were affected, there would be little concern about a chemical's neurotoxicity. If several unrelated neurologic functions were affected, but only at high doses and in conjunction with other overt signs of toxicity, including death or debilitation, there would be more concern. If several related functions were affected and the

effects appeared to be dose- and time-dependent, there would be still more concern. As the number of chemicals tested for neurotoxic potential increases, many different combinations of affected functions will emerge. Deciding which combinations of positive findings indicate the need to continue testing will not be trivial.

From many standpoints, FOBs have shortcomings. Most of their observations are semiquantitative. The sensitivity, reliability, and reproducibility of some have not been well documented. This deficiency can often be overcome by using more quantitative methods. For example, extensor thrust and grip strength can be measured with simple devices that use a strain gauge (Cabe and Tilson, 1978; Meyer et al., 1979). Hindlimb splay can be measured by inking the animal's paws, dropping it onto a sheet of paper, and measuring the distance between the footprints (Edwards and Parker, 1977). Equilibrium and muscle coordination can be measured by the length of time a rat can maintain itself on a rotating rod (Bogo et al.,

TABLE 4-6 End Points That Might be Included in a Functional Observational Battery[a]

In Home Cage and Open Field	Manipulative	Physiologic
Posture[b](D)	Ease of removal(R)	Body temperature (I)
Convulsions and tremors[b](D)	Ease of handling(R)	Body weight (I)
Palpebral closure[b](D)	Palpebral closure[b](R)	
Lacrimation[b](R)	Approach response[b](R)	
Piloerection[b](Q)	Click response[b](R)	
Salivation[b](R)	Tail-pinch response[b](R)	
Vocalizations[b](Q)	Righting reflex(R)	
Rearing[b](C)	Landing foot splay(I)	
Urination[b](C)	Forelimb grip strength[b](I)	
Defecation[b](C)	Hindlimb grip strength(I)	
Gait[b](R)	Pupil response(Q)	
Arousal[b](R)		
Mobility(R)		
Stereotypy(D)		
Bizarre behavior		

[a]Type of data yielded: D, descriptive; R, rank-order, scalar; Q, quantal data; I, interval, continuous; C, count.
[b]Specified in EPA guidelines (1986).
Source: Moser (1989).

1981). Tremor can be characterized with automated devices (Gerhart et al., 1985; Newland, 1988). Many studies have shown that the observations commonly included in FOBs can detect the effects of known neurotoxicants (Haggerty, 1989; Kulig, 1989; Moser, 1989; O'Donoghue, 1989).

Motor Activity

Motor activity includes a broad class of behaviors involving coordinated participation of sensory, motor, and integrative processes (MacPhail et al., 1989). Motor activity has several advantages for testing: it is noninvasive; motivational procedures, such as food deprivation, are not needed to produce it; and its recording is usually automated, and that reduces experimenter-animal interactions (Maurissen and Mattsson, 1989). Many studies have shown that motor activity can be affected by psychoactive and neuroactive chemicals (Reiter and MacPhail, 1979; Tilson, 1987; MacPhail et al., 1989). MacPhail et al. (1989) evaluated motor-activity measures and found them highly consistent across replications. The sensitivity of motor-activity measures in detecting reproducible neurotoxic effects is generally comparable with that of more sophisticated measures of neurobehavioral function.

Although motor-activity measures are often used to identify neurotoxic chemicals, they have disadvantages, particularly with regard to specificity of adverse effects on the nervous system. It has been argued that the results of motor-activity tests alone lack specificity and do not often provide information useful for later testing or characterization (Maurissen and Mattsson, 1989). However, specificity might not be as important in hazard identification as consistency and sensitivity. Motor-activity measurements are commonly used in conjunction with FOBs.

Schedule-Controlled Operant Behavior

Schedule-controlled operant behavior (SCOB) involves the maintenance of behavior (performance) by intermittent reinforcement. Different patterns of behavior and response rates are controlled by the relationship between response and later reinforcement. SCOB affords a measure of learned behavior and is useful for studying chemical-induced effects on motor, sensory, and cognitive function.

The primary end points for evaluation are agent-induced changes in response rate or frequency and the temporal pattern of responding. Response rate is usually related to an objective response, such as a lever press or key peck, and differs according to the schedule of reinforcement. Response rates are expressed per unit of time. For some classes of chemicals, the direction of an effect on response rate can differ between low and high doses. Agent-induced changes in temporal pattern of responding can occur independently of changes in rate and require analysis of the distribution of responses relative to the reinforcement schedule.

SCOB has been used to study the effects of psychoactive drugs on behavior and is sensitive to many neurotoxicants, including methylmercury, solvents, pesticides, acrylamides, carbon monoxide, and organic and inorganic lead (see, for example, MacPhail, 1985; Tilson, 1987; and Rice, 1988). The experimental animal often serves as its own control, and the procedure provides an opportunity to study a few animals extensively over a relatively long period. SCOB typically requires motivational procedures, such as food deprivation; and training sessions are usually required to establish a stable baseline of responding. Because of its sensitivity to neuroactive chemicals, SCOB has greater potential for use in hazard characterization than in hazard identification.

Many of the behavioral tests, and particularly the FOB, have been developed and validated with well-characterized neurotoxicants. It is much easier to interpret the results of behavioral tests when a large body of information already exists on a particular substance; one can use this information to help interpret the results. It is much more difficult to interpret the results of screening tests on a new product on which very little information is available.

Specialized Tests of Neurologic Function

Neurotoxicants produce a wide array of functional deficits, including motor, sensory, and learning-memory dysfunction. Many procedures have been devised to assess relatively subtle changes in those functions; hence their applicability to hazard characterization. Specialized tests and agents that affect them have been reviewed recently (WHO, 1986; Tilson, 1987) and are discussed only briefly below.

Motor Function

Motor dysfunction is a common neurotoxic effect, and many different types of tests have been devised to measure time- and dose-dependent effects. Anger (1984) reported 14 motor effects of 89 substances, which could be classified into four categories: tremor, convulsions, weakness, and incoordination. Chemical-induced changes in motor function can be determined with relatively simple techniques and may be used as a component of an FOB.

Several procedures have been used to characterize chemical-induced motor dysfunction. An example has been described by Newland (1988), who trained squirrel monkeys to hold a bar within specified limits (i.e., displacement) to receive positive reinforcement. The bar was also attached to a rotary device, which allowed measurement of chemical-induced tremor. Spectral analysis was used to characterize the tremor, which was found to be similar to those seen in humans exposed to neurotoxicants or with such neurologic diseases as Parkinson's disease.

Incoordination and performance changes can be assessed with procedures that measure chemical-induced alterations in force (Fowler, 1987). Animals are trained to receive positive reinforcement by applying force to a fixed lever. Training can also include maintenance of an appropriate force for a specified period. The accuracy of performance is sensitive to many psychoactive drugs (Walker et al., 1981; Newland, 1988). Gait has been measured in rats under standardized conditions and can be a sensitive indication of specific damage to the basal ganglia and motor cortex (Hruska et al., 1979) as well as damage to the spinal cord and peripheral nervous system.

Procedures to characterize chemical-induced motor dysfunction have not been used extensively in neurotoxicology. Most of them require pre-exposure training (including alterations of motivational state) of experimental animals. However, such tests might be useful, inasmuch as similar procedures are often used in assessing humans.

Sensory Function

Alterations in sensory processes (e.g., paresthesias and visual or auditory impairments) are frequently reported signs or symptoms in humans exposed to toxicants (Anger, 1984). Several approaches have been devised to measure sensory deficits. Data from tests of sensory function must be interpreted within the context of changes in body weight and body temperature. Furthermore, many tests assess the behavioral response of an animal to a specific stimulus; the response is usually a motor movement that could be directly affected by chemical exposure. Thus, care must be taken to determine whether proper controls were included to eliminate the possibility that changes in response to a stimulus were related to agent-induced motor dysfunction.

Several testing procedures have been devised to screen for sensory deficits. Many rely on orientation or the response of an animal to a stimulus. Such tests are usually included in the FOB used in screening (e.g., tail-pinch or click responses). Responses are usually recorded as being either present, absent, or changed in magnitude (Moser, 1989; O'Donoghue, 1989). The tests would not be suitable to characterize chemical-induced changes in acuity or fields of perception. Sensory deficits are usually characterized with psychophysical methods, which study the relationship between the physical dimensions of a stimulus and the behavioral response it generates (Maurissen, 1988).

One approach to the characterization of sensory function involves the use of reflex-modification techniques (Crofton and Sheets, 1989). A stimulus of varying intensity is presented before a stimulus that elicits a defined sensorimotor reflex. If the time between the two stimuli is appropriate, the response to the eliciting stimulus can be significantly inhibited (i.e., prepulse inhibition). The observation of inhibition is contingent on the ability of the animal to perceive the presence of the first stimulus. Agent-induced changes in the frequency or threshold required to inhibit the reflex are taken as possible agent-induced changes in sensory function. Changes in the ability of the first stimulus to inhibit the reflex must be interpreted within the context of changes in the response to the eliciting stimulus, i.e., a sensory change is inferred primarily on the basis of an agent's ability to alter the degree of inhibition of the reflex in the absence of any related change in sensorimotor function. Control runs should be performed to determine the basal response without the initial stimulus. Prepulse inhibition has been used

only recently in neurotoxicology (Fechter and Young, 1983) and can be used to assess sensory function in humans, as well as in experimental animals.

Various behavioral procedures require that a learned response occur only in the presence of a specific stimulus. Chemical-induced changes in sensory function are determined by altering the physical characteristics of the stimulus (e.g., magnitude or frequency) and measuring the alteration in response rate or accuracy. In an example of the use of a discriminated conditional response to assess chemical-induced sensory dysfunction, Maurissen et al. (1983) trained monkeys to report the presence of a vibratory or electric stimulus applied to the fingertip. Repeated dosing with acrylamide produced a persistent decrease in vibration sensitivity; sensitivity to electric stimulation was unimpaired. That pattern of sensory dysfunction corresponded well to known sensory deficits in humans. Discriminated conditional response procedures have been used to assess the ototoxicity produced by toluene (Pryor et al., 1983) and the visual toxicity produced by acrylamide (Merigan et al., 1982).

Procedures to characterize chemical-induced sensory dysfunction have been used often in neurotoxicology. As in the case of most procedures designed to characterize toxicity, training and motivational factors can be confounding factors. Many tests designed for laboratory animals can be applied to humans.

Learning and Memory

Learning and memory disorders are neurotoxic effects of great importance. Impairment of memory is reported fairly often by adult humans as a consequence of toxic exposure. Behavioral deficits in children have been caused by lead exposure (Smith et al., 1989). And it is hypothesized (Calne et al., 1986) that chronic low-level exposure to toxic agents can have a role in the pathogenesis of senile dementia.

Learning is defined as a lasting change in behavior, memory as the persistence of learned behavior. Alterations in learning and memory must be inferred from changes in behavior. However, changes in learning and memory must be separated from other changes in behavior that do not involve cognitive or associative processes (e.g., motor function, sensory capabilities, and motivational factors), and an apparent toxicant-induced change in learning or memory should be demonstrated over a range of stimuli and conditions. Before it is concluded that a toxicant alters learning and memory, effects should be confirmed in a second learning procedure. It is well known that lesions in some regions of the brain can facilitate some types of learning by removing behavioral tendencies (e.g., inhibitory responses due to stress) that moderate the rate of learning under normal circumstances. A discussion of learning procedures and examples of chemicals that can affect learning and memory have appeared in recent reviews (Heise, 1984; WHO, 1986; Tilson, 1987; Peele, 1989).

One simple end-point procedure to measure in assessing learning and memory is habituation, which is defined as a gradual decrease in the magnitude or frequency of a response after repeated presentations of a stimulus. A toxicant can affect habituation by increasing or decreasing the number of stimulus presentations. An example of chemical-induced effects on habituation can be found in a study by Overstreet (1977), who reported that diisopropyl fluorophosphate (DFP), a choline acetyltransferase inhibitor, had no effect on the response to a novel stimulus; with repeated presentations of the stimulus, however, DFP-treated rats habituated slower than controls. Habituation is a very simple form of learning and would be perturbed by a number of chemical effects not related to learning.

A general approach to studying the effects of a chemical on learning and memory involves the pairing of a novel stimulus with a second stimulus that produces a known, observable, and quantifiable response (i.e., classical "Pavlovian" conditioning). The novel stimulus is known as the conditioned stimulus, and the second, eliciting stimulus the unconditioned stimulus. With repeated pairings of the two stimuli, the conditioned stimulus comes to elicit a response similar to the response to the unconditioned stimulus. The procedure has been used in behavioral pharmacology and, to a lesser extent, in neurotoxicology. Neurotoxicants that interfere with learning and memory would alter the number of presentations of the pair of stimuli required to produce conditioning or learning. Memory would be tested by determining how long after the last presentation of the two stimuli the conditioned stimulus would still elicit a response. For example, Yokel (1983) dosed rabbits repeatedly with aluminum and found that exposed rabbits learned the conditioned eyeblink response more slowly than controls; such effects on establishing the conditioned relationship between the two stimuli were seen in the absence of aluminum-induced alterations in sensitivity to the unconditioned stimulus or ability to respond. Other classically conditioned responses known to be affected by psychoactive or neurotoxic agents are the conditioned taste aversion (Riley and Tuck, 1985) and conditioned suppression (Chiba and Ando, 1976).

Other procedures use instrumental learning, which involves the pairing of a response with a stimulus that increases the probability of future response through reinforcement. Response rate can be increased by using positive reinforcement or removing negative reinforcement. Learning is usually assessed by determining the number of presentations or trials needed to produce a defined frequency of response. Memory can be defined specifically as the maintenance of a stated frequency of response after initial training. Neurotoxicants may adversely affect learning by increasing or decreasing the number of presentations required to achieve the designated criterion. Decrements in memory may be indicated by a decrease in the probability or frequency of a response at some time after initial training. Toxicant-induced changes in learning and memory should be interpreted within the context of possible toxicant-induced changes in sensory, motor, and motivational factors. An example of a test based on instrumental learning is the repeated-acquisition procedure. It requires that an animal learn how to solve a series of problems that vary from session to session. The results of the repeated-acquisition procedure are affected by carbon monoxide and microwave exposures (Schrot et al., 1980, 1984). Other examples of instrumental learning procedures used in neurotoxicology are passive and active avoidance, Y-image avoidance, spatial mazes (radial-arm maze), and delayed matching to sample (see Heise, 1984; WHO, 1986; Tilson, 1987).

One problem with almost all measures of learning and memory in animals is that it is sometimes difficult to extrapolate the procedures to human testing or to predict analogous effects in humans. However, some tasks, such as the conditioned eyeblink test and delayed matching to sample, can be easily adapted for use with humans.

Neurophysiologic Procedures

Several clinical neurophysiologic tests have been applied to exposed animals for diagnostic or prognostic evaluation of the nervous system. The tests are noninvasive or only minimally invasive, readily adapted to longitudinal studies, and capable of detecting and measuring neurologic manifestations of neurotoxicity, regardless of the initiating mechanism. When properly chosen, they can probe the functional status of particularly affected portions of neuronal networks (such as reflexes and evoked responses). Their

results are reproducible within the constraints of biologic variation and the skill of the experimenter. Although of considerable value in risk assessment, they are nonetheless post hoc studies that yield results that reflect varied and often unknown exposures. The use and limitations of the tests have recently been reviewed by LeQuesne (1987).

Under conditions of the same locus and mechanism of neurotoxic action (such as demyelination), neurophysiologic testing in experimental animals appears to produce reliable indicators of what might be expected after similar exposure of other species, including humans. That can be illustrated by a brief comparison of data from clinical and experimental studies of acrylamide neurotoxicity. Clinical features of the neuropathy included ataxia, diminished or lost tendon reflexes, and sensory complaints (LeQuesne, 1980). Nerve-conduction velocities in affected patients were normal or (in a very few patients) just below normal (Takahashi et al., 1971). Studies in baboons revealed a similar picture of weakness and ataxia, also with minimal decrements in conduction velocities (Hopkins and Gilliatt, 1971). Cats with acrylamide intoxication had normal sensory and motor nerve-conduction velocities (Lowndes and Baker, 1976; Lowndes et al., 1978a), but were markedly ataxic and hyporeflexic. The close temporal association between onset and severity of signs of intoxication and impairment of static and dynamic properties of muscle spindles (Lowndes et al., 1978b) revealed that proprioceptive defects form the basis for the loss of tendon reflexes and almost certainly contribute to the ataxia.

Nerve-Conduction Studies

A number of end points can be recorded, but the critical variables are velocity of nerve conduction (e.g., in meters per second), response amplitude, and excitability, which is usually a measure of the time required before a nerve can generate another impulse (refractory period). Any long-term change in peripheral nerve function is likely to be accompanied by lesions observable with a light microscope. Although chemical-induced changes in the velocity of nerve conduction are generally rare, they are most likely to occur with decreases in body temperature and are associated with severe behavioral dysfunction. Changes in amplitude are often observed after exposure to neurotoxicants and can be associated with sensory or motor deficits. Changes in excitability are likely to cause altered sensory thresholds, altered behavioral reaction times, and altered susceptibility to seizures. Nerve-conduction responses depend on temperature, and studies should control for this variable.

The utility of maximal motor or sensory nerve-conduction velocities depends on the neuropathologic expression of neurotoxicity. Exposure resulting in demyelination markedly reduces maximal conduction velocity. In carbon disulfide neurotoxicity, sensory conduction velocities declined earlier than motor conduction velocities (Vasilescu, 1976). Similarly, neuropathies resulting from γ-diketone precursors (*n*-hexane and methyl-*n*-butylketone) and lead produce paranodal or segmental demyelination that is manifest in diminished conduction velocities (Spencer et al., 1975; Korobkin et al., 1975; Buchthal and Behse, 1979). Toxic neuropathies with axonal degeneration as hallmarks exhibit mild or no decrease in maximal conduction velocities; neuropathies resulting from γ-diketone precursors (LeQuesne and McLeod, 1977), organophosphates (Hierons and Johnson, 1978; Shiraishi et al., 1983), or acrylamide (Fullerton, 1969; Takahashi et al., 1971) are in this category, as are mild neuropathies caused by methylbutylketone (Allen et al., 1975).

Electrodiagnostic tests that rely on amplitudes of compound action potentials (muscle or sensory nerve action potentials) provide sensitive indexes of peripheral nerve abnormality. The tests depend on the number and synchrony of conducted impulses, which are reduced when there is a loss of contributing fibers or a dispersion due to conduction slowing. Sensory nerve action potentials are the more widely studied and are reportedly decreased in intoxication with acrylamide (Takahashi et al., 1971), Dipterex (Shiraishi et al., 1983), or methylmercury (Murai et al., 1982).

Sensory Evoked Potentials

Evoked-potential methods assess the neurophysiologic response of a particular sensory system to a particular stimulus (e.g., light or sound) and can help to detect and characterize neurotoxicity (Mattsson et al., 1989). The most commonly measured end points are peak latency (time from the onset of the stimulus to the peak response) and peak amplitude. Changes in latency imply changes in conduction velocity or in synaptic transmission. Acute exposures are likely to change latencies by changing body temperature. Long-term changes in latency might actually reflect alterations in conduction, which could be produced by alterations in myelination. When such changes are interpreted to reflect central, rather than peripheral, nervous system involvement, they might indicate an irreversible effect.

Changes in amplitude of peaks can imply changes in the number of nerve cells. Several peaks are averaged across numerous trials (stimulus presentations), and a change in average peak magnitude might be associated with an increase in variability of measurement. The importance of an amplitude change depends on the particular peak and sensory system under investigation. Changes in the evoked potential are likely to be associated with agent-induced sensory changes, which can be confirmed at the behavioral level.

Sensory evoked-potential techniques are used extensively in human neurotoxicity studies and are adapted easily to laboratory animals. The biologic basis of the electro-physiologic response is generally well understood and can be collected relatively quickly (Mattsson et al., 1989). The utility of sensory evoked potentials in neurotoxicity evaluation has been demonstrated, for example, in the case of toluene-induced auditory impairment (Rebert et al., 1983; Mattsson and Albee, 1988) and hexachlorophene-induced somatosensory dysfunction (Mattsson et al., 1989). However, the time and expertise required to implant electrodes and interpret the data and the recording equipment required reduce the likelihood that this technique will be used routinely for identification of chemical hazards.

Electroencephalography

Electroencephalographic (EEG) analysis is used widely in clinical settings for the diagnosis of neurologic disorders and less often for the detection of subtle toxicity-induced dysfunction. However, it is well known that dissociation between the EEG pattern and behavior can occur, particularly in chemical-treated animals. Changes in the pattern of the EEG can be elicited by stimuli that produce arousal (e.g., light and sound), by normal sleep, and by anesthetic drugs. In studies with toxicants, changes in EEG pattern can precede alterations in other objective signs of neurotoxicity. Experiments with the EEG must be done under highly controlled conditions, and the data must be considered case by case. EEG abnormalities have been noted in patients exposed to carbon disulfide (Seppalainen and Haltia, 1980) and alkylmetals (Cossa et al., 1959; Fortemps et al., 1978).

Neurochemical Procedures

Functions within the nervous system depend on synthesis of, release of, and receptor activation by specific neurotransmit-ters among specific groups of neurons. Many neurochemical end points could be measured in neurotoxicologic studies, including effects on neurotransmitters (e.g., changes in synthesis, transport, storage, release, re-uptake, or degradation of serotonin, norepinephrine, acetylcholine, or amino acids) and their receptors; effects on lipids, glycolipids, glycoproteins, or other constituents of neural membranes; effects on ion channels or membrane-bound enzymes that regulate neuronal activity; and effects on metabolic processes necessary to maintain neural activity (e.g., changes in glycogen and creatine phosphate concentrations, glucose availability, and mitochondrial structure and function). The large number of possible neurochemical effects is not amenable to the development of a battery of neurochemical tests to screen large numbers of chemicals. Results of in vivo tests at the behavioral or neurophysiologic level can suggest mechanistic hypotheses to test with biochemical and neurochemical end points (Mailman, 1987).

Neurochemical effects of neurotoxicants have been investigated in many laboratories (WHO, 1986; Mailman, 1987). Those of lead, for example, have been extensively studied in animals (Silbergeld and Hruska, 1980; Winder and Kitchen, 1984). Attempts have been made to use the results of these studies to further the understanding of lead-induced neurotoxicity in humans. For example, Silbergeld and Chisolm (1976) studied monoamine metabolites in urine of children exposed to lead and found a correlation between blood lead content and 24-hour urinary excretion of the dopamine metabolite homovanillic acid (HVA). HVA was measured before initiation of chelation therapy and within a week after the children had been removed from lead-contaminated environments. Over the long term, urinary HVA content was reduced, as was blood lead content, although both blood lead and urinary HVA remained higher in treated lead-exposed children than in age-matched controls. In contrast with dopamine, some neurotransmitters reported to be altered by lead in animal models—such as GABA and enkephalin—are less amenable to clinical measurement, because they require an invasive procedure to collect cerebrospinal fluid.

Neurochemical procedures appear to have great potential in development of biologic markers of neurotoxicant-induced exposure. The organophosphorus (OP) compounds have stimulated development of such biologic markers; an erythrocyte acetylcholinesterase assay has been used to monitor human exposure to OP compounds that cause cholinergic poisoning, and assays of lymphocyte neurotoxic esterase (NTE) activity have been used to detect exposure to OP chemicals that cause delayed neurotoxicity.

Recently, hemoglobin adducts (such as the lysyl amino groups of hemoglobin modified by γ-diketones) have been evaluated as possible biologic markers of exposure to some neurotoxicants. Because of the long lifetime of the erythrocyte (120 days in humans) and because erythrocytes become more dense as they age, density-gradient centrifugation of blood samples allows differentiation between recent and earlier exposure to the neurotoxicant. The same sample of blood can be used to detect protein cross-linking, because spectrin dimers eluted from membranes of lysed erythrocytes will be stable if cross-linking occurred in vivo (St. Clair et al., 1988).

A variety of neurochemical probes of neuronal and glial integrity could be used to evaluate the possible neurotoxic effects of a candidate chemical in experimental animals (O'Callaghan, 1988). As discussed with regard to in vitro systems, these neurochemical assessments could deal with very general indicators, such as total protein or DNA, in a given region of the nervous system. Alternatively, more specific indicators—such as transmitter-related enzymes, various receptors, or the state of phosphorylation of axonal proteins—could be compared

in control and experimental groups of animals. One much-used assay detects glial fibrillary acidic protein (GFAP). The immunoassay for GFAP revealed areas of neuronal damage detectable with silver stain but not with routine neuropathologic studies. Agents and conditions that increase brain GFAP include trimethyltin, triethyltin, methylmercury, cadmium, MPTP, stab wounds, and aging. The use of GFAP as a sensitive and specific markers of central neuronal damage is simple and provides quantitative data that can be useful in risk assessment. The choice of indicators should be made in the light of any structure-activity relationship or other information that suggests an optimal test strategy for a given agent.

Neuroendocrine Interactions

Neuroendocrine toxicology is an emerging field. Despite the relative paucity of formal studies, suspicion that toxicants can affect neuroendocrine processes has been voiced since the earliest case studies of poisoning. For example, reports of lead poisoning in historical times comment on the short stature of children exposed to lead (Schwartz et al., 1986). Similar observations have been made in cases of infantile methylmercury intoxication (Matsumoto et al., 1965; Takeuchi, 1977). Anecdotal reports suggesting neuroendocrine toxicity have appeared in conjunction with pesticides, drugs of abuse, medicinal agents, and industrial residues. However, there has been no systematic review of neuroendocrine toxicity in conjunction with any of those classes of chemicals.

Neuroendocrine toxicology should receive greater attention than it does now. Evidence is growing that endocrine organs are important targets for many toxicants and that deleterious effects have not been recognized, simply because no one has looked for them. For instance, prenatal exposure to TCDD at very low concentrations alters sexual behaviors in male rats by an endocrine-dependent mechanism related to early hormone imprinting of the CNS (Moore et al., 1991). It is important to note that, because damage to the neuroendocrine system might be expressed as dysfunction or injury at the target or end organ, the neural contribution to the effect could be misinterpreted. For example, some chemicals described as reproductive toxicants exert their primary effect on the pituitary-reproductive axis (NRC, 1989b). Similarly, growth effects of teratogens can result from injury to the hypothalamic neurons that control release of growth hormone from the pituitary, rather than from direct effects of the agents on somatic tissues (Rodier et al., 1991).

Many areas of the brain responsible for regulating hormone production or specific hormonal response are very small and specialized. Groups of only several thousand or even hundreds of cells can be responsible for major endocrine functions. Although the size of those specialized regions in the brain makes them difficult to identify during routine histologic assessments, their dysfunction can have major consequences. The regulation of the endocrine system has been well characterized, so routine techniques are available to identify endocrine perturbations and to pinpoint their sites of origin; these should be more widely used.

The ability to detect disturbances in the neuroendocrine axis by provocative challenge, by radioimmunoassays of plasma from peripheral blood, or by end-organ evaluation has implications for risk assessment. The techniques are somewhat less invasive than autopsy or brain biopsy, are more quantitative than many behavioral procedures, and (under optimal circumstances) can pinpoint sites of neurotoxic injury with a high degree of specificity. The neuroendocrine system is an interactive axis involving many structures, and a toxicant can act on end organ, brain, or both.

Neuropathology

Neurologic lesions can be classified according to their characteristics or site of action. Lesions can be classified as neuronopathies (changes in the neuronal body), axonopathies (changes in the axons), myelinopathies (changes in the myelin sheaths), dendropathies (changes in the dendrites), and peripheral neuropathies (changes in the peripheral nerves). For axonopathies, a more precise location of the changes should be described (i.e., proximal or distal; central or peripheral).

In general, chemical effects lead to two general types of primary alterations: the accumulation, proliferation, or rearrangement of structural elements (e.g., intermediate filaments and microtubules) or organelles (e.g., mitochondria); and the breakdown of cells, in whole or in part. Partial cellular breakdown can be associated with regenerative processes that can follow chemical exposure.

Most neurotoxic damage is evident at the microscopic level, but gross changes in structure can be reflected in significant changes in the weight of the brain. Weight changes, discoloration, discrete or massive hemorrhage, and obvious lesions are clear indicators of adverse effects.

Careful histologic analyses form the cornerstone of our knowledge of toxic neuropathies that alter nervous tissue structurally, and the changes can be unambiguously correlated with clinical outcomes and with aging. Many structural changes have been identified neurohistologically. Methylmercury produces selective granule cell loss in the cerebellum (Hunter and Russell, 1954) and axonal degeneration in sensory ganglion cells (Cavanaugh and Chen, 1971). Degeneration of the long sensory and motor axons in both the central and peripheral nervous systems is produced by OP compounds, acrylamide, and 2,5-hexanedione (Cavanaugh, 1964; Spencer and Schaumburg, 1977a,b). Myelin loss has been seen as a result of diphtheria with no axon loss (Cavanaugh and Jacobs, 1964) and as a result of exposure to lead with only slight loss (Fullerton, 1966); myelin vacuolization in the CNS and PNS has been caused by hexachlorophene (Towfighi et al., 1974) and in central neurons by triethyltin (Aleu et al., 1963).

A careful examination of properly preserved and prepared tissues (which are generally available only from experimental animals) establishes or rules out structural damage, identifies the most vulnerable sites within the nervous system, and traces the temporal evolution of pathologic changes. The nervous system can be systematically assessed by sampling its various areas or particular tissues, as recommended by different expert groups (Tables 4-7 and 4-8). Table 4-7 reflects a recommendation by the World Health Organization (1986), and Table 4-8 contains a somewhat expanded sample of tissues believed adequate to identify most known disease states and vulnerable areas of the nervous system with light microscopy (Spencer et al., 1980). As the number of anatomic regions assessed increases, so do the labor and resource requirements of the assays. Sampling all regions listed in Tables 4-7 and 4-8 could be quite costly. Fuller characterization of tissue damage and identification of mechanisms of effect at the structural level often require electron microscopy guided by light microscopy (WHO, 1986).

Imaging Procedures

Available neuroimaging procedures to assess structural changes in the brain can be divided into three major categories: computed axial tomography (CAT), magnetic resonance imaging (MRI) and nuclear magnetic resonance spectroscopy (MRS), and positron-emission tomography (PET).

The first two categories of neuroimaging procedures can be carried out with current clinical techniques. The CAT scan combines computerized imaging techniques with radiologic tomography to develop a detailed image of the brain. Because the technique is noninvasive, it has gained widespread clinical acceptance in the last 15 years. Contrast agents can be used to enhance visualization of some abnormalities, such as tumors or vascular malformations.

The newer MRI technique is similar to CAT, but, rather than using ionizing radiation, recreates anatomic information with nuclear magnetic moments. It requires a high-intensity magnetic field, and it monitors changes in nuclear magnetic moment relaxation times after the magnetic field has been applied. MRI is unique in that it uses three kinds of data to characterize anatomic information: nuclear-specific data, i.e., data specific to the nuclear magnetic moment; imaging-specific data, i.e., data specific to the way MRI is being performed; and tissue-specific data, i.e., data specific to characteristics of

TABLE 4-7 Areas of the Nervous System to be Used in Neuropathologic Evaluation

Primary Areas	Secondary Areas
Cerebellum	Olfactory epithelium and tubercles
Brain stem	Inner ear and labyrinths
Pituitary gland	Plantar nerves and skin receptors
Eye, oculomotor muscles, optic nerve	Autonomic ganglia
Spinal cord	Nerves and organs of innervations
Sensory ganglia	
Sciatic nerve (branches from vertebral column to ankles and selected muscles it innervates)	

Adapted from WHO (1986).

TABLE 4-8   Tissues of the Nervous System to be Used in Neuropathologic Evaluation

Central Tissues	Peripheral Tissues
Motor cortex	Gasserian ganglion
Visual cortex	Lumbar dorsal root ganglia
Subfornical organ	Lumbar dorsal root
Area postrema	Lumbar ventral root
Lateral geniculate body	Proximal sciatic nerve
Optic tract	Tibial nerve at knee
Optic nerve	Tibial nerve, calf muscle branches
Retina	Plantar nerves at ankle
Cerebellar vermis	Gastrocnemius muscle
Gracile nucleus	Lumbrical muscle spindles
Cuneate nucleus	Lumbrical neuromuscular junctions
Gracile tract (T6, L5)	
Ventromedial tract (medulla oblongata, T6, L5)	
Dorsal spinocerebellar tract	
(medulla oblongata)	
Hypoglossal nucleus	
Descending tract of V	
Lumbar cord, anterior horn	
Mammillary bodies	
Hypothalamus	
Hippocampus	
Striatum	
Substantia nigra	

Adapted from Spencer et al. (1980).

the tissues that are being imaged. The three kinds of data provide some of the unusual power of MRI to define precisely the pathophysiology of CNS disease. The procedure also provides a detailed image of the brain and might be even safer than CAT, in that there is no exposure to radiation.

The use of nuclear MRS can provide a more detailed evaluation of the biochemical status of the CNS. The kinds of biochemical information obtained from nuclear MRS include rates of energy generation and use, substrate metabolism, blood flow, and tissue integrity. Taken together, the uniqueness of the anatomic and biochemical data provided by MRI and MRS suggests that the combination of the two techniques will become a powerful adjunct to neurotoxicology both in humans and in experimental animals.

PET is potentially much more sensitive.

It represents a major advance in the ability to monitor some types of neurotoxicity by labeling compounds with positron-emitting elements (e.g., manganese and fluorine) so that it is possible to observe alterations in neurochemistry, as well as in receptors in the brain of living subjects. Before the introduction of PET scanning, such examination was possible only in postmortem material.

The potential value of PET was recently recognized in the study of a group of heroin addicts exposed to MPTP. A number of addicts mistakenly took synthetic heroin contaminated with the dopaminergic neurotoxicant MPTP; some became severely parkinsonian, but many did not. By using positron-emitting labeled 6-fluorodopa, striatal dopamine deficiency was shown in those who were exposed to MPTP, but were not symptomatic (Calne et al., 1985). That finding has already led to the hope that similar techniques will make it possible to predict idiopathic Parkinson's disease.

Another PET scanning technique involves the evaluation of cerebral glucose use. Scanning revealed decreased glucose metabolism in the brains of people poisoned by ingesting mussels contaminated with domoic acid (Teitelbaum et al., 1990). In patients with Alzheimer's disease, some areas of the cerebral cortex use less glucose (Friedland et al., 1989). Toxicants, whether their effects are selective or diffuse, are likely to decrease oxygen consumption and glucose use at their target sites. That might provide a way of monitoring the physiologic consequences of exposure to a wide variety of neurotoxicants.

The potential applications of PET procedures are enormous. If a neurotoxicant damages one or more areas of the brain, causing a change in neurotransmitter concentrations, it might be possible to measure the effects of the toxicant early, while they are still subclinical. The techniques can be used in experimental animals for longitudinal studies, because it is not necessary to kill them at different times to follow the evolution of neurotoxic effects. Ligands have been developed to image various receptor populations throughout the brain. Thus, it could eventually be possible to examine the effects of neurotoxicants on a wide variety of neurotransmitter systems and their corresponding receptor populations in living subjects.

Both CAT and MRI techniques can be used to demonstrate gross structural alterations in the brain. They might be useful in defining any neurotoxic process that causes substantial atrophy in the CNS. A specific example is the cerebellar and cortical atrophy that can occur after exposure to organic solvents—glue sniffer's encephalopathy. The limitation of the procedures is that they require major, if not massive, loss of neuronal populations before atrophy is visible. Other clinical measures would probably demonstrate abnormalities

much earlier in the course of neurotoxic exposure.

Imaging procedures are not useful for assessing spinal-cord, nerve, and muscle damage. They are also expensive, and there is only a small data base available on their use in neurotoxicology.

Modifying Variables

Not all people exposed to a given neurotoxicant respond similarly. The determinants of susceptibility are not always well understood. Some variables, however, are known to modify responses to neurotoxic agents and must be taken into account in screening and any testing at higher tiers. The importance of age at exposure and at testing has already been mentioned in Chapter 2. This section indicates how age, sex, and genetic background of subjects should be considered in neurotoxicity testing.

Age at Exposure

Organisms can differ in sensitivity to neurotoxicants at different ages. Developing organs, for example, resist damage from low oxygen concentrations (a condition that is normal in utero). But, because of their high mitotic activity, they are more sensitive to antimitotic agents, such as x rays. Some

TABLE 4-9   Tests Used in NCTR Collaborative Study

Auditory-startle habituation
1-hour activity in figure-8 maze
Diurnal activity in figure-8 maze
Activity before and after amphetamine challenge
Visual-discrimination learning
Negative geotaxis
Olfactory discrimination

Source: Buelke-Sam et al. (1985).

agents (such as diazepam) cause temporary effects in the mature CNS, but permanent effects in the developing CNS (Kellogg, 1985); others (such as methylmercury) are toxic at every age, but toxic at lower doses in developing organisms (Spyker, 1975). Some differential effects can be explained on the basis of pharmacokinetics, but that is not the case for any of the agents discussed above. Thus, it is never safe to assume that effects observed in adults can be simply extrapolated to developing subjects. Developing organisms are not always more sensitive to toxic agents than adults; they are sometimes qualitatively different in their responses, because they are biologically different.

The difference between developing animals' and mature animals' responses to toxicity in other body systems is already recognized in regulations that require developmental exposure as a separate part of toxicity testing. The teratology guidelines focus on malformations and provide no information on the integrity of the nervous system when injuries are less obvious than anencephaly or exencephaly; greater attention to neurologic effects after developmental exposures is needed. Guidelines being written for the Environmental Protection Agency are based on recommendations from several groups and on experience with various test batteries.

Butcher and Vorhees (1979) proposed a neurobehavioral test battery for developing animals. Tests were selected to reflect the guidelines for assessing reproductive toxicity then in force in Japan, Britain, and France. The battery was validated with vitamin A, a classic behavioral teratogen (Butcher and Vorhees, 1979). The Butcher and Vorhees battery was later expanded by Vorhees and others to include tests on pivoting, olfactory orientation, auditory startle, neonatal T maze, figure-8 activity, M-shaped water maze, open-field grip strength, midair righting, rotorod, crossing parallel rods, jumping down to home cage, elevated maze

learning, and E-shaped water maze (Adams, 1986).

The National Center for Toxicological Research (NCTR), which serves as a research arm of the Food and Drug Administration (FDA), conducted a major project known as the NCTR Collaborative Study. It was designed to demonstrate the reliability of classical behavioral-teratology tests (Table 4-9) when administered in five laboratories. The tests proved to be highly reliable between and within laboratories, and the effects of methylmercuric chloride on auditory startle were identified in all laboratories (Buelke-Sam et al., 1985).

Developmental neurotoxicology is a relatively new and important field. Because functional development of offspring can be influenced by maternal toxicity or chemical-induced growth deficits, care must be taken in the interpretation of data from developmental-neurotoxicity studies. Whether developmental neurotoxicologic assessments should be done at the hazard-identification or hazard-detection phase is being debated. Certainly concerns raised about the ability of a chemical to affect reproduction and development would apply to the postnatal growth and function of organisms exposed during development.

Age at Testing

Age at testing is an important modifying variable, for several reasons. When injury occurs early in life, it might damage systems that are not fully mature and thus not be fully expressed until later in life. The classic demonstrations of that phenomenon are in monkeys with lesions, in which some deficits take as long as 2 years to express their full effect (Goldman, 1971). It is not a characteristic of the damage itself, but of the organism on which it is imposed. Thus, just as antimitotic injury from radiation to the hippocampus during gestation will not lead to hyperactivity until after puberty (Rodier et al., 1975), a mechanical lesion of the hippocampus will have a delayed effect (Isaacson et al., 1968). The decrease in norepinephrine content and turnover in the hypothalamus that follows fetal exposure to diazepam is minor at 30 days in the rat, when the norepinephrine system is immature, but greater than 50% at 90 days, when the system is mature (Kellogg, 1985).

A separate phenomenon is the increasing effect of lesions with aging—initially suggested to describe performance after neonatal x irradiation, which differed from performance in controls in preweaning animals, was near normal in young adults, and then differed again as animals aged (Wallace et al., 1972). Brain weights of treated subjects were always low, but the difference from control subjects became more pronounced with advancing age. The same pattern is suspected to account for the late onset of parkinsonian dementia after early exposure to cycad products. It is possible that a natural decline in function uncovers a longstanding deficit.

It is not easy to introduce aging as a variable in screening tests, because lifetime studies in animals are expensive. More basic research is needed to determine how often a toxic insult interacts with aging. In the meantime, it is reasonable to hold test animals until maturity, rather than doing short-term studies. Although short-term studies might be appropriate for teratology experiments aimed at gross structural changes, they preclude evaluation of normal adult function of the nervous system.

Sex

Toxicity to sexually dimorphic structures can be sex-specific, and the brain is in some respects a sexually dimorphic structure. A few examples of how the effects of some toxic agents on the nervous system might depend on sex are known, although not necessarily understood. Developmental effects of methylmercury on motor tests were more pronounced in boys than in girls in studies of a fish-eating population (McKeown-Eyssen et al., 1983), and neonatal male mice show more mitotic arrest and later cerebellar loss in response to methylmercury than do females, even though they do not have higher concentrations in the brain after being treated (Sager et al., 1984). Some sex differences might be due to pharmacokinetic differences, rather than differences in the target tissue, but that does not weaken the point that it is reasonable to test both sexes for effects.

Genetic Differences

Specific genetic differences in response to toxic agents have not been commonly demonstrated, and without some knowledge of mechanisms it is unlikely that susceptible populations can be identified in the screening of chemicals. However, when such populations are known, they should be evaluated. The classic human example is the case of the recessive gene for metabolism of the amino acid phenylalanine. People who are homozygous for the normal allele can clear excess phenylalanine; those who are homozygous for the mutant are severely brain-damaged by postnatal exposure to the amounts of phenylalanine found in a normal diet. Heterozygotes fall between the extremes in their metabolic capacity and should be sensitive to high concentrations of phenylalanine, but not harmed by the amount in a normal diet. Some toxicologists have been concerned about exposure of this group to the phenylalanine sweetener aspartame. Because most heterozygotes are unaware of their status with regard to this gene, they cannot avoid products that might be hazardous to them.

CURRENT REGULATORY APPROACHES

The Organization for Economic Cooperation and Development (OECD) (Brydon et al., 1990) and other organizations (Tilson, 1990b) have recommended simple and rapid observations of neurobehavioral function of experimental animals in their home cages or in large open fields in protocols for many subchronic and chronic tests. Such observations include those of naturally occurring behaviors (e.g., rearing, stereotypy, and urination) and procedurally easy and quick assessments (e.g., of click response, landing foot splay, and righting reflex) (Moser, 1989). Other home-cage testing protocols involve more elaborate 24-hour recording of naturally occurring behaviors (Evans, 1989). It is not clear how effective such relatively general and nonbinding instructions have been in fostering the detection of neurotoxicity in the context of more general toxicity testing. Moreover, such approaches do not provide for the detection of alterations in complex behaviors, such as learning and motor performance, inasmuch as their testing requires special instruments and, depending on the test, training of the animals.

No U.S. federal laws are directed specifically at protecting humans from systemic toxicity, much less from neurotoxic hazards specifically; but regulatory agencies have considerable latitude to emphasize specific health effects (Fisher, 1980; Reiter, 1987; Sette, 1987; OTA, 1990). With the exception of a test for delayed neuropathy in hens required by EPA for OP esters (Federal Register, 1978), there are no specific regulatory requirements to evaluate neurotoxicity in any current federal guidelines or regulations.

The testing requirements for neurotoxicity vary among the regulatory agencies. Premarket testing is required for pesticides by EPA under the Federal Insecticide, Fungicide, and Rodenticide Act (FIFRA) and for drugs and food additives by FDA, but not specifically testing for neurotoxicity. The Consumer Product Safety Commission and EPA may require testing of consumer products and industrial chemicals, respectively, but only if they can justify the need for such testing. In practice, neurotoxicity testing is rarely required, but is usually left to the discretion of the manufacturer.

Until passage of the Toxic Substances Control Act (TSCA) in 1976, there was no legal mechanism in the United States for prospective evaluation of the neurotoxicity of new industrial chemicals (OTA, 1990). Under TSCA (Section 5), there must be some indication that a "new" chemical might be neurotoxic before EPA can require specific tests in the premanufacturing notice (PMN) program, but then EPA can prohibit the chemical from entering into commerce until the required data are available. Generally, determination of whether additional testing is needed is based on a structure-activity analysis, because about 50% of PMNs are submitted with no toxicity data and the toxicity data that are submitted are rarely more than the results of acute lethality or irritation tests. Of the 6,120 PMNs submitted in 1984-1987, 1,200 underwent additional review (about half these were found to be associated with unreasonable risk); of those identified for additional review, 220 were suspected of being potentially neurotoxic (OTA, 1990). The SAR-based approach is now regarded as generally insensitive and problematic. Moreover, many thousands of potentially neurotoxic compounds developed before passage of TSCA remain untested. Except in unusual cases, the chemicals were simply registered with EPA and added to the TSCA inventory. For "old" chemicals (TSCA, Section 4), EPA can issue a test rule and require testing only if neurotoxicity is suspected, but little progress has been made in establishing final requirements for testing such compounds or completing evaluation of what new data have been generated (GAO, 1990a,b). Proposed test rules or consent decrees for 19 chemicals or chemical classes have included requirements for neurotoxicity testing (OTA, 1990). FIFRA also has a provision for the retroactive reassessment of the toxicity of active ingredients in existing pesticides, but this program is progressing slowly (OTA, 1990).

For assessing suspected neurotoxicants, EPA's Office of Toxic Substances has issued a series of test guidelines (Federal Register, 1985). They include guidelines for testing motor activity and a functional observational battery. Schedule-controlled operant behavior, neuropathology, peripheral nerve function, and organophosphate-induced delayed neuropathy are also addressed. A testing guideline for developmental neurotoxicity is pending. Those guidelines are generally intended to be generic guides as to how to make particular measurements, rather than lists of specific techniques. Data generated with the guidelines are used in conjunction with other toxicity data to establish potential risks and to establish reference doses.

FDA follows a pattern similar to that of EPA for assessing potential neurotoxicity. Therapeutic agents are assessed in preclinical (animal) and clinical (human) studies requiring careful, but unspecified, observations (Fisher, 1980). Specific neurotoxicity tests are required only for drugs that are intended to be neuroactive; for other drugs, a review for neurotoxicity is undertaken only if such effects are suspected (OTA, 1990). FDA requires neurotoxicity testing of food additives only if neurotoxicity is suspected for a particular chemical on the basis of evidence developed in traditional toxicity testing. However, FDA requires testing when a stated "level of concern" is reached on the basis of anticipated exposure, potential toxicity, or structure-activity relationships. A general histopathologic assessment of several sections of the brain, spinal cord, and peripheral nerve is required, with notation of any observed behavioral abnormalities in the test animals. Positive evidence that is revealed by the screening tests might, at the agency's option, lead to further assessments,

whose nature is decided case by case. In fact, only instances of neuropathy or clinical signs of neurotoxicity—such as paralysis, tremor, or convulsions—have an impact on FDA decisions (Sobotka, 1986).

Rodent reproductive studies, which might permit the observation of developmental problems, are ordinarily conducted for drugs that will be administered to women of childbearing age (OTA, 1990). On the whole, however, current tests for developmental toxicity (particularly at FDA) are designed primarily to detect structural malformations or fetal death, which are appraised after sacrifice before delivery. Even in the short-term Chernoff-Kavlock protocol (1982) and in the standard protocols for reproductive and developmental effects, the pups are sacrificed before they acquire much behavior, so opportunities for observation of behavioral teratogenicity and other developmental deficits are not provided.

Other countries have explicitly prescribed behavioral tests in animals to screen for developmental neurotoxic effects of new drugs that possibly will be used by pregnant women. To allow a new drug on the market, Japan requires a fertility study, a teratology study with female exposures, and a perinatal study with exposures during the last third of gestation and through lactation. In the latter two studies, some offspring are examined for "development (including behavioral development)." Great Britain also provides general specifications, requiring tests for "auditory, visual, and behavioral impairment" for premarket screening of new drugs. The European Economic Community guidelines require assessment of "auditory, visual and behavioral impairment" (Vorhees, 1986).

The testing protocol of the National Toxicology Program (NTP) provides for fairly extensive neuropathologic assessment, but relatively little functional observation. NTP tests have detected the neurotoxic effects of sodium azide, for instance, because it caused neuronal destruction in chronically treated rodents. However, it is generally accepted that finding mild neuronal degeneration, at the light microscopic level, in the absence of functional or biochemical information, is unlikely to be a sensitive means of detecting neurotoxicity.

STRATEGIES FOR IMPROVED NEUROTOXICITY TESTING

Screening

Several current proposals for generalized approaches to neurotoxicity testing of new and existing chemicals recommend a three-tiered scheme (Moser, 1989; Tilson, 1990a; OTA, 1990). The first tier (the screen) consists of tests for the existence of a neuro-toxic hazard. If no evidence of such a haz-ard is found within the first tier, toxicity testing ceases. Positive findings in the first tier would terminate the development of many chemicals being investigated for com-mercial applications, raising the dilemma of false positives discussed later. For other apparently neurotoxic chemicals (those already in use and those whose commercial development is nonetheless deemed promi-sing), neurotoxicity testing would continue to the second tier.

The experiments undertaken in the second tier would characterize the nature of a substance's neurotoxicity, e.g., its specific target tissue and dose-response and time-response relationships. The selection of specific experiments in the second tier would be directed by the results of earlier testing. Preliminary second-tier dose-response assays would establish the potency of the substance for the adverse neurotoxic end points iden-tified in the first tier. Comparison with the chemical's potency for other types of toxicity and with estimated magnitudes of human exposure is likely to determine whether extensive additional effort is put into charac-terizing its neurotoxicity or whether it pro-gresses to the third tier for investigation of its mechanisms of action in detail. If other

effects would predominate or no neurologic consequences were likely to ensue from plausible exposures, research resources would be better directed toward other ques-tions.

Testing in the third tier would aim to specify the mechanism by which the chemical produces its neurotoxic effects. This type of detailed investigation would be appropriate for substances with broad exposure that produce serious effects (e.g., lead), substan-ces with low or rare exposure that produce a particular noteworthy effect (e.g., MPTP), or series of substances that produce lesions of concern (e.g., demyelination). The infor-mation obtained in the investigations would be of use in attempting to control the most serious existing exposure problems and in establishing SARs for the more efficient prediction of neurotoxicity of chemicals considered in the future.

Despite an apparent simplicity in the design of such a tiered strategy, implemen-tation is not easy; many questions remain. Determination of the component tests of the first tier is the most crucial aspect of the three-tiered approach, because truly positive substances will not continue to later tiers unless detected at this point, and the course of substances that are studied further will tend to be tailored to their performance in earlier tests. The first tier is the heart of the screening aspect of neurotoxicity testing; the later tiers might produce data of great value in advancing understanding of neurotoxic processes and how the nervous system oper-ates, but the first tier is the first line in preventing neurotoxic disease.

How are chemicals ranked for entry into the first tier? Does the first tier represent a single battery of tests to which all entering chemicals will be subjected, or is the flow of testing in the first tier itself directed by the profile of results? One way to proceed is to have an initial tier composed of a number of tests that are applied to all candidate sub-stances. Substances that do not show signs of neurotoxicity are not tested further. Confidence in such a battery is a function of the number of tests in it, the quality of the tests, the size of the experiments (number of subjects, doses, durations of exposure, etc.), and the degree to which the tests examine a wide range of potential neurotoxic outcomes (learning, memory, developmental effects, etc.) or other biologic markers that may serve as surrogate measures for them. The sensitivity of a neurotoxicologic screen might be increased by including additional tests. Many of the tests were mentioned earlier. Inclusion of some of them might increase confidence in screening procedures, but at the price of consuming more resources. That suggests a tradeoff between testing an individual substance thoroughly and processing more substances. Tests that might be considered for addition to the screen in the future would have to be validated and their incremental value demonstrated.

The neurotoxicologic literature has many references to two-tiered and multitiered testing. In many instances, what is being discussed is the multistage nature of complete neurotoxic testing, i.e., the validation of the screening decision and the progression from screening (hazard identification) to experimental neurotoxic tests designed to establish quantitative dose-response relationships, mechanisms of action, etc. However, other authors appear to be discussing "multistaged screening" in the strict sense of testing for the purpose of hazard identification (tier one). If all steps of screening are required for every substance, then the plan is no different from a proposal for a greatly expanded comprehensive screening battery. However, if multistaged testing is required of only some substances, then it is important to specify the criteria for deciding which substances should pass beyond the initial level. Requiring additional levels or types of testing for substances in some chemical classes is, in effect, merely adding more tests to the first tier.

Some type of functional observational battery (FOB) has most often been proposed for the screening phase of neurotoxicologic assessment. It appears to be the simplest,

most comprehensive option available and it, at a minimum, should be included in the first tier of neurotoxicity testing. Depending on the particular test, behavioral end points are often used in the detection phase of testing, because they may be global indicators of many of the sensory, motor, and integrative processes of the central and peripheral nervous systems. Observational batteries have been used in rodents by many investigators to assess chemicals for neurotoxicity and have been recommended by several expert panels (see Tilson, 1990b). The EPA Office of Toxic Substances prepared guidelines (EPA, 1985) for testing potential neurotoxicants that included an FOB consisting of several measures that are relatively simple, noninvasive, and quick to perform. A description of the FOB and a summary of responses produced by representative neurotoxicants were published recently by Moser and colleagues (Moser et al., 1988; Moser, 1989). Provision should be made for acute and repeated dosing and for observation of the animals over an extended period. It would also be desirable to measure motor activity and perform limited neuropathologic tests on the treated animals, as recommended by EPA (1985). Suggestive screening findings or structural alerts derived from knowledge of SARs should invoke a second level of tests, in which impaired acquisition of complex behaviors after exposure during development is used as a general indication of neurodevelopmental toxicity. In addition to tests covering the various aspects of neurotoxic effects of concern for every chemical, a screening battery might contain specific procedures used only for particular classes of chemicals, as is the case for testing OP pesticides for possible delayed distal neuropathy.

Validation

The procedures just described are the only general neurotoxicity methods whose validation approaches what would be necessary for a routinely applied screen. Before they could be adopted, additional or other tests would have to be thoroughly appraised for reproducibility and for relevance to human health. Their success in distinguishing the properties of an extensive battery of nonneurotoxicants and substances that act by a full spectrum of mechanisms to produce a given end point would have to be demonstrated. Establishing a control battery of appropriate chemicals is in itself a challenge, given the fairly limited extent to which the universe of chemicals has been tested for neurotoxicity. Ever-increasing experience might ultimately permit the substitution of a battery of short-term in vitro assays for longer procedures in whole animals. In addition to its screening capacity, such an in vitro battery would be expected to yield considerably more information about mechanisms of action than the whole-animal approach to screening with an FOB and thereby allow much more directed testing in the later tiers.

Priority-Setting and Implementation

The proposed implementation of large-scale screening for neurotoxicity has generated controversy about the appropriate strategy for approaching the first tier of testing, hazard identification. It is generally agreed that substances already present in the environment for which there is any indication of neurotoxicity in the form of structural alerts or preliminary observations should be evaluated thoroughly. Similarly, in the case of chemicals being considered for future use, such signals of possible neurotoxicity should trigger rigorous testing, if commercial development is to be pursued. When there are few or no data on which to base a judgment as to whether a substance (new or in use) presents a neurotoxic hazard, the estimated volume of production and expected pattern of human exposure (concentrations, frequen-

cies, and numbers of people) are the most reasonable bases on which to set priorities and begin screening.

As reviewed by the Office of Technology Assessment (1990), federal regulations provide the regulatory agencies with the authority to demand that toxicity testing, including tests for neurotoxicity, be performed on chemicals as a requirement for registration or to establish standards for continued use. Federal agencies have not exercised that authority often, nor followed through completely in analyzing the data that have been generated (GAO, 1990a,b). Although it might seem obvious that the sponsors of an innovative product should be responsible for testing it adequately, the question remains of who should be responsible for conducting and financing the necessary testing on substances already widely present in the environment. It would be desirable if the burden of neurotoxicity screening not only could be divided and coordinated among U.S. industry, government, and academe, but also could involve an international effort, as is being encouraged by OECD (Brydon et al., 1990) for general toxicity testing.

Research Needs

An important question is how in vitro systems can contribute importantly to the neurotoxicologic evaluation of a broad range of agents. Research is needed to determine whether culture systems can reliably identify known neurotoxic agents and active related compounds and distinguish them from related, but inactive (or less active) compounds. Research is also needed to develop in vitro techniques that can identify neurotoxicants that require metabolic activation to be effective and those which do not, that affect either adult or developing organisms, and that are injurious after either acute or chronic exposures.

Empirical results of the many in vitro systems available need to be correlated with

TABLE 4-10 Proposed Components for Evaluating In Vitro Neurotoxicity Screening Tests

Neurotoxicant		Test System	
Site of Action	Positive Chemical Controls	Culture System	Test Measures
Excitable membranes	Pyrethroids, ouabain	Cell lines: PC12 C-6	Neuronal survival: cell counts tetanus toxin binding
Transmitter systems	Anticholinesterases, 6-hydroxydopamine	Dissociated cell cultures: cerebral cortex spinal cord neurons glia	Glia: glial fibrillary acidic protein
Axons	Acrylamide, 2,5-hexanedione, vincristine	Cell cultures: sympathetic ganglion	Myelination: structure biochemical measures
Dendrites	Excitatory amino acids	Reaggregates: basal ganglia	Transmitter-related: choline acetyltransferase glutamic acid decarboxylase aminergic enzymes transmitter uptake agonist binding
Soma	Trimethyltin, doxorubicin (Adriamycin), ricin	Explants: dorsal root ganglia and spinal cord, cerebellum, cerebral cortex	Structural changes
Astroglia	Mercury	Special sensory: retina organ of Corti	Excitable membranes: sodium uptake calcium uptake saxitoxin binding

studies in whole animals (indeed, human populations) to establish the neurotoxicologic implications of findings from in vitro preparations. Substantial efforts will be needed to provide funding and organizational resources for that type of work; but the work is essential if reliable, efficient test systems are to be developed reasonably soon. In the long term, mechanistic studies are needed to increase understanding of the fundamental molecular targets of neurotoxic chemicals. Various schemes can be used to outline a more general approach (Table 4-10). Accumulated knowledge should eventually allow relatively efficient evaluation of agents that are likely to fall in one or another general category of neurotoxicant. A general model has emerged from developmental neurobiology that might similarly allow focusing of neurotoxicologic research with in vitro systems that capture one or another step in the progression from developing embryo to degenerating adult nervous system (Figure 4-1).

Routinized screening and mechanistic studies will produce a growing body of detailed neurotoxicologic data on diverse chemicals that should be intensively analyzed. Ultimately, study of the data should reveal SARs that are more generally useful than the few that are now considered well established and that will be much better substantiated and more reliable than most current conjectures.

Much of the controversy over proper testing procedures arises from two intrinsically conflicting objectives: minimizing the incidence of false positives (substances are incorrectly identified as hazardous) and minimizing the incidence of false negatives (substances are incorrectly identified as nonhazardous). Both goals are desirable, but they cannot be maximized simultaneously. Too high an incidence of false positives

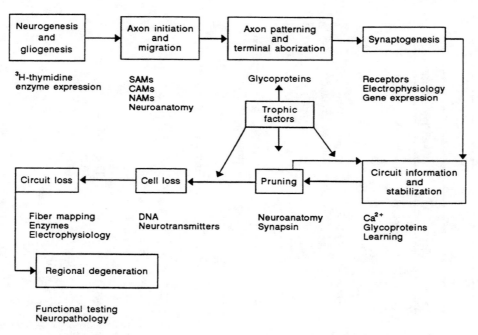

FIGURE 4-1   Biologic markers in the stages between formation and degeneration of neural circuits. Selected markers are listed under the stages that are named in boxes.

will waste our resources, worsen the already severe logistic problem of testing thousands of chemicals, and cause at least some potentially valuable chemicals to be discarded. In the screening situation, however, a high incidence of false negatives is probably more undesirable. Failure to detect a truly neuro-toxic substance will expose our society to health hazards, with the potential for tragic consequences like those associated with one false negative, thalidomide. Testing for characterization should expose false posi-tives; the abandonment of an occasional new chemical on the basis of what are actually false-positive screening results is the likely cost of this process.

The expense and effort of neurotoxico-logic screening would be very great if it were undertaken in isolation. Considerable prac-tical advantages would accrue if neurotox-icologic screening were integrated with a multidisciplinary program of general toxico-logic and teratologic screening.

5

Surveillance to Prevent
Neurotoxicity in Humans

Need for Surveillance

Neurologic disease and dysfunction after exposure to toxic chemicals in the environment could be prevented at least in part through premanufacturing and premarket screening (Chapter 4) of all newly synthesized chemicals, coupled with retroactive screening of chemical substances already in commerce. Such screening is intended as primary prevention; it is meant to prevent neurologic disease and dysfunction of environmental origin by identifying new neurotoxicants before they enter the environment and before human exposure has occurred and then handling them in such a way that human exposure will be as close to zero as possible.

Current American programs for detecting neurotoxicants, including industrial chemicals, through premarket testing have serious deficiencies (NRC, 1984). A major problem is that many potentially neurotoxic compounds that came into use before passage of the Toxic Substances Control Act (TSCA) in 1976 remain untested and are still not required to be tested. Current neurotoxicity testing procedures developed under the TSCA mandate have not been standardized for new industrial chemicals (other than

pesticides) and are not sensitive enough to detect the effects of chronic low-dose exposures. Neurotoxins in natural products are most likely to escape testing and control, in part because of confusion as to which federal agencies are responsible for this control. Primary prevention is far from perfect; it has sometimes failed in the past, and it cannot be expected to be totally successful in the future.

The early identification of disease or dysfunction and acting on the identification constitute secondary prevention. Its goals are to prevent progression of disease in identified cases, to treat early enough to cure if that is possible, and to prevent the occurrence of additional cases by providing a basis for primary prevention. The methods of postmarketing secondary prevention of environmental neurotoxic disease include environmental monitoring and surveillance of the human population to detect neurotoxic disorders as early as possible, which are combined with efforts to keep potentially toxic exposures to a minimum. Epidemiologic studies and clinical screening are complementary facets of maintaining alertness so that neurotoxic illness will not go unnoticed. Secondary prevention constitutes a necessary and critical backup to primary prevention.

The training of physicians, particularly industrial physicians, to recognize neurotoxic manifestations should substantially improve secondary prevention efforts.

Secondary prevention is not a substitute for premarket testing of chemicals for potential neurotoxicity; rather, the data from a secondary prevention program complement and supplement the data derived from premarket laboratory testing. Although toxicologic information derived from future in vitro and in vivo investigations might be expected to form an ever-increasing proportion of the neurotoxicity data base, responsibility to conduct epidemiologic surveillance, to monitor toxic exposures, and to study incidents of human exposure to neurotoxicants when they occur will remain.

Surveillance and monitoring play oversight or exploratory roles. Fundamentally, they should be viewed as tools for identifying possible problems-hypothesis-generating exercises. In contrast, classical epidemiologic studies of populations exposed to neurotoxicants (e.g., Needleman, 1986), clinical trials of pharmaceuticals, and occasionally laboratory experiments on humans (e.g., Dick and Johnson, 1986; Johnson, 1987) are hypothesis-testing and will not be considered here in detail.

Effective surveillance systems are characterized by systematic and rapid collection of data, efficient analysis and evaluation of data, and prompt communication of the results obtained. In discussing general principles of health surveillance, Foege et al. (1976) noted that dissemination of a program's information often enhances cooperation in contributing more information to the system.

Surveillance of occupational and environmental health can be divided broadly into systems for monitoring of exposure (to detect hazards) and monitoring of disease (to detect health responses). Hazard surveillance systems monitor the use, storage, and environmental release of chemical toxicants and tabulate the numbers of people exposed to them. Their purpose might be to detect instances of poor industrial hygiene, inappropriate or excessive use of pharmaceuticals, or improper waste disposal or to generate exposure data to complement epidemiologic and clinical data on toxic end points. Health surveillance systems record the occurrence of designated diseases. They might have the general population as their target or focus on specific high-risk groups (Baker, 1989; Landrigan, 1989). Both types of surveillance are necessary, and both are practiced by public-health agencies in the United States and elsewhere; the data generated by the two types of systems are complementary. Surveillance systems can be categorized according to their purpose: to monitor the handling of known neurotoxicants, to follow the effect of control measures, or to detect previously unrecognized problems.

Recognizing a previously unrecognized problem is most difficult, because no system can be expected to find reliably what is not specifically sought or to engender confidence that something not specifically looked for is really not present. Because the measurement of environmental concentrations of chemicals requires discrete decisions about what substances will be sampled and where and when, it is clear what information has and has not been sought. Medical monitoring can be somewhat more open-ended, and that raises the possibility of uncertainties about the definitiveness of negative findings; but the storage and retrieval of observations necessitate specification of format and limitations on the type of information accepted. In addition, having a data base truly useful for quantitative estimation of occurrence rates demands that the means by which the data were ascertained be clearly defined, so that it is known what the sample represents.

A perennial problem in epidemiologic surveillance is that information on the baseline incidence and prevalence of neurologic diseases is sparse. Thus, it is difficult to detect anything but large outbreaks or clusters of neurologic disease. The development of a substantial data set would require significant resources. Much knowledge about neurotoxicants and neurotoxic disease has, however, come from observation of humans—from case reports to large-scale epidemiologic studies. Most reports of outbreaks of neurologic diseases have come from medical clinics, state health departments, industrial medical groups, or consultative programs, such as the Health Hazard Evaluation Program of the National Institute for Occupational Safety and Health (NIOSH). Outbreaks have all too often served as late-stage sentinel events identifying the neurotoxicity of chemical substances. Many times, apparent excesses or clusters of disease have, on close examination, proved to be false leads.

NEUROBEHAVIORAL TEST BATTERIES

In Chapter 1 (Table 1-1), the great diversity of forms that neurotoxicity can take was demonstrated by the extensive list of neurologic signs and symptoms attributed to exposure to various chemicals. Table 5-1 shows that a temporal pattern of exposure and response can characterize particular chemicals. A complete neurologic examination and occupational history (Table 5-2) are essential for identifying specific neurologic syndromes (Johnson, 1987). The examination summarized in Table 5-2 is typically supplemented by tests of CNS function, which can be as simple as asking the subject to name the current president of the United States or asking the subject at the end of the examination for three words (e.g., hat, dog, and tree) that were presented at the beginning of the examination. More sophisticated neuropsychologic tests can be used when there is reason to suspect CNS problems.

Testing human subjects for neurotoxic injury combines clinicians' efforts to identify the cause of diseases or disorders and neuropsychologists' or experimental psychologists' use of techniques to assess performance. Field investigators might adopt a unique battery of tests for each chemical to be assessed or select from among the 5-10 standardized batteries now in use. If latitude is possible in selecting tests, selection is based on signs reported in the exposed group and the known neurotoxic effects of the chemical under study (or of structurally related chemicals). That general approach has led to the use of hundreds of tests in various worksite studies (Johnson and Anger, 1983; Anger, 1990) and has typified NIOSH research (Anger, 1985). Several human neurotoxicity test batteries have been developed and are being validated. Otto and Eckerman (1985), Anger (1990), and Johnson et al. (1990) have described the major batteries, including several implemented on computers.

Finland's Institute of Occupational Health pioneered neurotoxicity testing at the worksite in the 1950s. Investigators at the institute developed a test battery well adapted to monitoring the main concerns in Finnish industry (primarily solvents). Their tests (Table 5-3), which have been optimized by factor analysis over the years, appraise various psychologic domains (Hänninen and Lindström, 1979). The battery is now used in neurotoxicity evaluations of several worker groups in Finland, including prospective studies on new workers.

An apical test battery (Table 5-3) based on information-processing theory (Marteniuk, 1976) has been developed and tested in the field in Australia (Williamson et al., 1982; Williamson, 1990). For example, battery and lead-smelter workers with blood lead concentrations of 1.2-3.9 μmol/L (25-81 μg/dL) demonstrated poorer performance on all tests than a set of matched controls (Williamson and Teo, 1986).

At a 1983 meeting sponsored by the World Health Organization (WHO), an international team of experts on testing for

TABLE 5-1 Characteristics of Responses to Exposure to Some Neurotoxicants

Neurotoxicant	Exposure Pattern	Susceptible Development Stage	Reversibility of Response	Neurologic Category of Outcome Neurotoxicity	Typical
Acrylamide	Chronic	Late	Reversible, irreversible	Sensory, motor	Perepheral neuropathy, permanent visual loss, reversible hypoesthesia
n-Hexane	Chronic	Late	Reversible in early stages	Sensory, motor	Peripheral neuropathy
Carbon disulfide	Chronic	Late	Reversible in early stages, irreversible	Psychologic, cognitive, motor, sensory	Subjective complaints, visual loss, peripheral neuropathy
MPTP	Acute, chronic	Late	Irreversible, delayed	Motor	Parkinsonian syndrome
Methanol	Acute	Late	Irreversible	Sensory	Blindness from retinal damage, basal ganglia damage
Ethanol	Chronic	Prenatal	Irreversible	Cognitive, motor	Fetal alcohol syndrome

Organophosphates	Chronic	Late	Reversible, irreversible	Psychologic, motor, sensory	Acute dysfunction, long-term neuropathy, spinal-cord damage
Lead	Chronic	Childhood	Irreversible	Cognitive, motor	Reduced scores on developmental tests, peripheral nueropathy, encephalopathy
Mercury	Chronic	Early	Reversible	Psychologic, motor	Erethism, tremor
Methylmercury	Chronic	Early	Irreversible	Cognitive, motor	Visual, sensory, and motor dysfunction; retarded development
Manganese	Chronic	Late	Irreversible	Motor	Dystonia, behavioral aberrationsk extrapyramidal dysfunction
Chlorpromazine	Chronic	Late	Irreversible	Motor	Tardive dyskinesia
Vincristine	Chronic	Late	Reversible	Sensory, motor	Paresthesia, weakness, peripheral and cranial neuropathy

COLORADO COLLEGE LIBRARY
COLORADO SPRINGS
COLORADO

TABLE 5-2   Components of Clinical Neurologic Examination

General appearance
    Tremor—arms outstretched and at rest
    Ability to sit still
    Speech—clear, slurred
    Gums, nails, color of skin

Vital signs

Cranial nerves
    I   Sense of smell tested
    II   Funduscopic examination, disk margins, visual acuity, visual fields
    III   Pupil size and reactivity to light; extraocular movements—nystagmus
    IV   Extraocular movements
    V   Pin and touch over face, corneal reflex
    VI   Extraocular movements
    VII   Symmetry of facial movement
    VIII Hearing acuity, vestibular function
    IX-X Presence of gag reflex, ability to swallow
    XI   Symmetry of shoulder bulk and movement
    XII   Tongue: midline presence of atrophy or abnormal movements

Motor examination
    Presence of atrophy, fasciculations, tone—resistance to passive movement
    Ability to hold arms outstretched with eyes closed
    Grip strength, scored on scale of 0 (absence of any movement) to 5 (normal strength)
    Deep knee bend
    Hopping on each foot, walking on heels and toes
    Finger-tapping
    Extensor plantar response

Tendon reflexes, scored as 0 (absent), 1 (decreased), 2 (normal), 3 (increased), 4 (grossly exaggerated)
    Biceps
    Triceps
    Knee
    Ankle jerk

Coordination
    Tandem gait
    Finger-to-nose pointing
    Rapid alternating movement
    Foot-tapping

Gait
    Stance, presence of arm swing, rapid turning
    Ability to walk on toes and heels
    Walk quickly
    Run

Sensory
    Vibration and pin-testing in arms and legs
    Position sensation

Source: Johnson (1987).

neurotoxic effects in humans recommended the Neurobehavioral Core Test Battery (NCTB). The NCTB (Table 5-3) can be used to identify a broad range of neurotoxic effects. The primary goal was to generate uniform, more consistent data from a broad spectrum of occupations and neurotoxic exposure conditions. Tests included in the set had been used successfully in worksite studies (they had identified group differences associated with chemical exposures) and were believed to reflect a wide range of functional areas in humans. A supplemental series of tests was specified for further characterization (Johnson, 1987).

The Neurobehavioral Evaluation System (NES), available on IBM PC and portable Compaq computers, combines several neurobehavioral tests that have been used successfully in clinical settings or in field studies (Table 5-3). Tests are occasionally added to the battery, which is following an evolutionary course dictated by current interest in the field. The NES includes variants of five of the seven WHO NCTB tests, and the developers of the NES recommend that the user select tasks to be used according to specific exposure situations (Baker et al., 1985). That is, the battery provides a menu of tests without an explicit decision strategy for selecting among them for screening or other purposes. The NES has been used in more laboratory and field studies than any other battery in the last 5 years (Letz, 1991).

Anger (1989) matched the components of the WHO NCTB, Finnish, and NES batteries to the neurobehavioral effects that he had found to be reported most commonly after chemical exposures (Table 1-1, Chapter 1). He concluded that the NCTB would be the most comprehensive in detecting those effects. The Finnish battery lacks the ability to detect affective changes, and the NES does not have an established test of motor function. Although cognitive and, to a smaller degree, motor effects should be recognized fairly well by all three batteries, they all have limited potential for detecting

sensory deficits. All three batteries include tests to assess the more subtle CNS deficits produced by lower exposures, which are the subject of increasing concern.

The tests noted above have been developed primarily for screening, but other tests have been used to characterize the effects found in the screen. They are not fundamentally different, but add information about the nature of the problem. They include laboratory-based behavioral tests, clinical neuropsychologic examinations, and imaging evaluations. Some neurologic and behavioral tests also have been adapted to the monitoring for early identification of disease or indicators of neurotoxicity. Screening and monitoring tests have been developed to identify peripheral problems (such as toxic distal axonopathies) resulting from exposure to n-hexane, methyl n-butylketone, acrylamide, and related neurotoxicants. The neurologic examination often identifies neurotoxicity at an early, but clinically observable, stage. Specific monitoring or screening devices (e.g., the Optacon and the Vibratron II for assessment of vibratory sensitivity and NTE II to detect losses in temperature sensitivity) have successfully identified neuropathy at the preclinical stage (Arezzo and Schaumburg, 1980, 1989). A more specific test has been used to assess vibratory sensitivity in monkeys, and it is now used in the workplace to identify peripheral neuropathy (Arezzo et al., 1983).

Tests of auditory threshold have been used to detect hearing loss after exposures to noise and to drugs that affect the auditory system (e.g., quinine, arsenic, and glycoside antibiotics) (Stebbins et al., 1987). Similarly, visual function has been found to be impaired in animals and people exposed to methylmercury (Evans et al., 1975). Each of those tests has been used to identify early sensory loss so that exposures can be stopped before the development of irreversible effects.

Although there is debate about some screening tests for field applications, the

TABLE 5-3   Test Batteries

A. Finland Institute of Occupational Health Test Battery

Benton visual-retention test	Reaction time
Bourdon-Wiersma	Santa Ana
Symmetry drawing	Wechsler Memory Scale (portions)
Mira test	Wechsler Adult Intelligence Scale (portions)

   Source: Hänninen and Lindstrom (1979).

B. Australian Battery Based on Information-Processing Theory

Critical flicker fusion	Sensory store (iconic) memory
Vigilance	Sternberg Memory Test
Simple reaction time	Paired associates
Visual pursuit	Short-term memory
Hand steadiness	Long-term memory

   Source: Williamson et al. (1982).

C. World Health Organization Neurobehavioral Core Test Battery (NCTB)

Santa Ana	Benton visual-retention test
Aiming motor	Digit span
Simple reaction time	Profile of mood states
Digit symbol	

   Source: Johnson (1987).

D. Computer-Administered Neurobehavioral Evaluation System (NES)

Psychomotor performance	Memory and learning
Digit symbol[a]	Digit span[a]
Hand-eye coordination	Paired-associate learning
Simple reaction time[a]	Paired-associate recall
Continuous-performance test	Visual retention[a]
Finger-tapping	Pattern memory
	Memory scanning
Perceptual ability	Serial digit learning
Pattern comparison	
	Cognitive
Affect	Vocabulary
Mood test[a]	Horizontal addition
	Switching attention

   Source: Adapted from Letz and Baker (1986).

[a]Variant of WHO core test (Battery C)

tests can identify some forms of peripheral neuropathy when they are still reversible. Some companies test routinely to monitor for neuropathy affecting the sensory axons—a kind of secondary prevention system. Of course, such strategies cannot be applied to the testing of newly developed chemicals, to chemicals with effects that are not clearly defined, or to situations in which it is not known what chemical, if any, could cause any health effect.

It can be difficult to interpret the results of test batteries. On the whole, they do not yet have established norms based on extensive population testing, but they do provide objective measures suited to test-retest (before-and-after) assessments. Baseline (pre-exposure) performance data have not been routinely collected from workers before exposure, so unexposed people must be used for comparison. These performance effects are now usually evaluated not by self-matching or against established standards, but by comparison to a referent or control group.

The immense diversity of behavior makes it difficult for a simple test battery to screen adequately in a comprehensive manner; a battery cannot detect an effect for which it lacks a test. Although it is not likely that an entirely new type of health effect will be generated by a new chemical or class of chemical, it is likely that tests for additional end points will be needed as more chemicals are tested. As understanding of the various mechanisms of neurotoxicity increases, confidence that a comprehensive battery can be assembled should grow.

The established batteries of human tests are clearly limited in their screening capacity. No comprehensive rationale for test selection has been posited. The Australians Williamson et al. (1982) developed their screening battery around cognitive information theory, but this theory is now dated and is insufficiently related to medical approaches to neurologic evaluation or to neuropsychologic approaches. Eckerman et al. (1985) proposed a battery in which the tests are selected to assess eight cognitive functions identified by factor analysis; this battery has not been used in worksite evaluations. The field needs to develop a comprehensive rationale for the development of a neurotoxicity screening battery. Ideally, it would be based on the functions of the nervous system and the range of potential neurotoxic effects, and that is not yet possible.

In the absence of a comprehensive rationale for the selection of tests to compose a neurotoxicity screening battery, a good rationale would be to select tests that are sensitive to the range of neurotoxic effects that have been identified. In a comprehensive review of the literature, Anger and Johnson (1985) identified more than 120 neurotoxic effects. The range of neurotoxic effects might be far too broad to test comprehensively. Anger (1986) reviewed the list of 120 effects and identified the 35 reported as occurring after exposure to 25 or more chemicals (Table 1-1). The result formed a basis for selecting the most important functions to assess, given current knowledge. None of the batteries assesses sensory effects thoroughly, and the computer-based NES does not use a proven test of motor performance. However, the NES and some other batteries have the advantage of being administered by computer, which reduces the costs of administering the tests substantially. Some of the most common neurotoxic effects—particularly some forms of peripheral neuropathy and affective symptoms, weakness, ataxia, and many sensory effects—would not be detected by any of the batteries. (Of course, CNS changes can be correlated with some of these effects, in which case these batteries would be performing their function of detecting health effects. As noted earlier, a screening battery is developed for detection, not for characterization.)

The WHO NCTB contributors first suggested human neurotoxicity testing based on the core-test idea, which uses a standard set of tests in all studies, supplemented by additional tests when appropriate (this is a variant of the tier-testing idea) and is consistent with the idea of evaluating the most common health effects. The NES can be used to test most of the same functions tested by the WHO NCTB, and new tests are added periodically. That will allow adaptation to new problems as they arise. The tests of the NES are largely cognitive and so will approach the complex functions that are not identifiable without sophisticated tests, although tests of motor and sensory function are certainly required for truly comprehensive assessment. Both the NES and the NCTB have been used in different countries and languages; that fosters the development of an international data base that will allow exchange of test results.

Although those test batteries appear to be best suited to assessments of neurotoxic exposures in humans, their usefulness needs constant examination in light of new types of neurotoxicity, possible changes in the most frequently reported health effects (as chemical use shifts), and new cognitive or neurobiologic theories of nervous system function. That is, NES and NCTB batteries are the best choice today, but this judgment is based on identified health effects, rather than on understanding of the nervous system and how it functions. Fundamental progress will be based on advances in basic and theoretical neurobiology; data from continued testing of exposed population can be expected to lead only to minor improvements in test selection. An intermediate goal is the collection of data with a small number of reliable and valid test batteries, to provide a solid base of relatable data on various chemicals. Such a collection could lead to the development of specific criteria for prediction of effects (most desirably on the basis of chemical structure) or the development of a better screening battery.

CURRENT EXPOSURE SURVEILLANCE EFFORTS

Several sources of information provide data on potential for occupational exposure to neurotoxic substances in the United States. However, none of them is comprehensive, represents a well-defined worker population or industrial setting, or corresponds to any existing set of information on adverse health effects (NRC, 1987a).

The Integrated Management Information System (IMIS) of the Occupational Safety and Health Administration (OSHA) contains information obtained as a result of inspections for accidents or complaints, followups on violations, and general scheduled inspections. The only exposure data systematically entered into the IMIS are data obtained through analysis of environmental samples taken at worksites where citations were issued. Because IMIS samples are not random samples, generalizable information on the distribution of neurotoxic occupational exposures does not exist.

Industry-generated data constitute a second source of information on neurotoxic hazards. For 11 toxic substances (including lead) for which OSHA has set standards, industry is required to maintain records of exposure. However, because the records are not part of a central data system, such as the system maintained by OSHA, they are most useful for monitoring specific industries and employers and are far less useful for surveillance and discovery.

NIOSH conducts health hazard evaluations (HHEs) at the request of employers, employees, or their representatives. At times, both exposure monitoring and biologic sampling are done. The records provide insight into particular occupations and industries, but, like the OSHA data, they are not readily generalizable. NIOSH has conducted two national surveys of chemical exposure in nonagricultural businesses: the National Occupational Hazard Survey (NOHS) from 1972 to 1974 and the National

Occupational Exposure Survey (NOES) from 1981 to 1983. Each entailed walk-through inspections of approximately 5,000 plants, and potential exposures were tabulated by trained observers. However, no industrial-hygiene samples were collected. As a result, data on actual exposures are not available. The NOHS-NOES data have been used to identify general industrial groups in which neurotoxic exposures may be anticipated, and they have served as a valuable resource for epidemiologists and public-health officials (ASPH, 1988). Information on exposure of workers and consumers to pesticides is available from the Environmental Protection Agency (EPA).

Evidence of exposures to toxic substances in the general environment might be detected by the National Human Adipose Tissue Survey (NHATS), periodic sample surveys done by the National Center for Health Statistics (the National Health and Nutrition Examination Surveys, or NHANES), and toxic inventories maintained by states and cities under right-to-know legislation. Those data bases provide information on a range of chemicals—some that are known to be neurotoxic (e.g., lead, surveyed under NHANES II [NCHS, 1984]), some that accumulate in fat (such as pesticides, surveyed in the NHATS), and some that are stored and released in substantial quantities in industrial and other sites. The National Research Council has published a review of the NHATS program and made several recommendations for its modification and improvement (NRC, 1991b).

A particularly valuable new source of information is the Toxic Release Inventory (TRI), compiled annually by EPA since 1988 to monitor more than 300 chemicals released by industry into the environment. The inventory is mandated under the Superfund Amendments and Reauthorization Act (SARA). Seventeen of the 25 toxic substances with the highest volumes of release into the environment have neurotoxic potential (OTA, 1990).

Efforts are also needed to improve the availability and use of exposure-surveillance data. IMIS, the OSHA inspection data base, would be more useful if it covered all states in a more uniform fashion and if it included a larger, more representative sample of exposure data. Other types of exposure-reporting systems (such as those under federal and state right-to-know laws) should be evaluated for their utility in surveillance. For the preponderance of industrial settings, adequate exposure data are not available. Such information should be collected according to explicit sampling plans with up-to-date measurement techniques and stored in accessible data bases. It could then be used to target efforts to control known dangerous exposures or matched to health-effects files in an effort to detect new problems, as discussed below. Exposure surveillance is a continuous effort both to monitor changes in exposure and to monitor new industries or new uses of well-known chemicals (Fowler and Silbergeld, 1989).

When the objective is to assess broad trends in population exposure to neurotoxicants and to gather information on groups potentially at risk, selection of an appropriate surveillance strategy will depend on the expected degree of toxicity of the chemicals under study, the size of the exposed population, and the intensity of exposures. In essence, hazard or exposure surveillance consists of obtaining information on the distribution and use of chemicals, on the distribution of exposures, and on the size and distribution of the exposed population.

CURRENT DISEASE-SURVEILLANCE EFFORTS

A disease-surveillance system directed toward detecting problems caused by unrecognized neurotoxicants must be much more open and versatile than a system intended only to tabulate the incidence and prevalence of and mortality from a known disease. Neurotoxic disorders are particularly difficult to monitor (NIOSH, 1986; Friedlander and Hearne, 1980), and present surveillance systems are not very effective. Neurotoxic disorders are easy to overlook or misdiagnose, because their signs and symptoms often develop slowly and subtly. And some can be reversible or self-limiting. The patterns of effects produced by many neurotoxic chemicals are similar, consisting of common, nonspecific complaints, often mimicking diseases with other etiologies. The association between an observed health effect and the culpable environmental agent is rarely obvious. Clinical tests used in diagnosis are usually too imprecise or insensitive to identify the full spectrum of adverse health effects. A latent period between exposure and overt response will further complicate determination of causation.

Several existing data systems have been suggested as potentially useful for surveillance for neurotoxic disease that might be attributable to environmental exposures (Gable, 1990). They include vital records, health surveys, specifically designed surveillance systems, and the U.S. Census.

Since 1971, the Bureau of Labor Statistics (BLS) has conducted the Annual Survey of Illness and Injury with a well-constructed (and well-defined) probability sample; the most serious problem with this data base for the detection of neurotoxic illness is that the occurrence of illnesses, especially those involving latent periods, is vastly underreported. BLS also maintains a Supplementary Data System in which data from the states' workers' compensation programs are compiled; the data gathered differ from state to state, and the lack of total reference populations precludes the calculation of rates.

The National Center for Health Statistics (NCHS) compiles data (from death certificates) on every death in the country through the National Death Index. The basic data on death certificates are fairly uniform across states, and demographic cause-of-death information and some occupational data are almost always available. Occupational data are of uneven quality, which usually depends on whether anyone uses them for research purposes. The categories "housewife" and "retired" are often reported in the place of out-of-home employment or one-time major employment. Data from the Bureau of the Census are appropriate as denominators in calculating rates. NCHS makes the encoded mortality data available to researchers on computer tapes. The National Death Index is valuable for determining the vital status of subjects in epidemiologic followup studies; for subjects who have died, researchers are directed to the appropriate state for detailed information. NCHS also collects data from the states on every birth and fetal death, including birthweight, Apgar scores, congenital anomalies, and complications of pregnancy or birth. The birth records are not as uniform across states as are death certificates, and only a few states record information on parental occupation. NCHS has conducted the National Health Interview Survey, a large (120,000 persons), stratified probability sample of the civilian, noninstitutionalized population, every year since 1957. In addition to standard demographic information, both occupational status and data on illnesses, injuries, disabilities, and use of medical services are gathered. The fourth NHANES, a probability sample of the civilian, noninstitutionalized population also conducted by NCHS, is currently being carried out on a full-scale basis (the first was carried out in 1970). Physical examinations, laboratory tests, and responses to a questionnaire yield a detailed medical picture, but occupational information was not gathered in the previous surveys. In the current survey, job histories are being requested, and neurotoxicity is among five work-related conditions about which information is sought.

Other data bases are of potential value for the detection of neurotoxic disease. Regional poison-control centers, insurance companies, the Social Security Administration, and the Department of Veterans Affairs might have information useful for a specific surveillance project, but most fall short by not having both health and occupational data (data on actual exposures are almost never found), by being drawn from an inadequately defined population, by having restricted availability, or by being limited to a period that is not of interest.

The National Research Council Panel on Occupational and Health Statistics (NRC, 1987a), after reviewing the various sources of information on adverse health events, concluded that the information available on occupational injuries is poor and that the information on occupational illnesses is considerably worse. Those data sources were not designed to determine the causes of disease, nor do they emphasize occupational or environmental exposures. Thus, the contribution of such exposures to the etiology of a given disease is not usually recognized (ASPH, 1988). The end points selected for surveillance are usually restricted to overt clinical disease or death.

Once cases are reported, officials can collect further information and provide appropriate followup and suggestions for prevention in the workplace. The reporting systems must be interactive and provide a means for followup of reported cases to ensure continued physician involvement in the surveillance program. State and local health departments, in collaboration with occupational clinic groups, probably provide the best basis for such reporting systems.

When an excess or cluster of a particular symptom associated with neurotoxicity (National Conference on Clustering of Health Events [1990]) or a new syndrome with involvement of the nervous system is detected, determination of the likely causes is the next stage of the epidemiologic process. It requires careful comparison between observed physical and temporal patterns of exposures to possible toxicants and the occurrence of the cases.

The NIOSH Health Hazard Evaluation Program and similar activities of academic or government groups provide mechanisms for rapid evaluation of reported outbreaks and can therefore confirm initial reports of a possible neurotoxic problem. California has pioneered the development of a reporting system for outbreaks of pesticide poisoning in which collaborating health and agricultural departments can intervene. However, similar systems for the reporting of such diseases by physicians are not well established elsewhere in the United States (ASPH, 1988).

The committee has considered several ideas for improving disease surveillance systems:

• *Training of medical professionals.* Many medical professionals who are in direct contact with persons with neurotoxic disorders of environmental origin lack the knowledge necessary to recognize these ailments or to identify their etiology. The ability to diagnose the disorders and to associate them with workplace or community exposures must be improved (U.S. House of Representatives, 1986). Physicians need to be made more aware that environmental exposures can produce toxic effects and should be trained to obtain an account of exposures received occupationally, avocationally, etc., as well as medically, when taking a history (Goldman and Peters, 1981). Better initial training in occupational and environmental medicine and the dissemination of more current information on neurotoxic illnesses would benefit both health practitioners and their patients (NRC, 1987a). The development of improved diagnostic tests and criteria would facilitate the recognition of an occurrence of neurotoxicity when it appears.

• *Standardized disease definitions.* Uniform clinical definitions of the disorders in question are needed to provide a common reporting basis for physicians. Such a nomenclature would also make the assembled data more amenable to analysis and interpretation. NIOSH has a project to develop definitions for a series of occupational disorders.

• *Sentinel health events.* A sentinel health event (SHE) is a preventable disease, disability, or untimely death whose occurrence serves as a warning that the quality of preventive or therapeutic medical care might need to be improved (Rutstein et al., 1976). The idea has been refined for application to occupational situations; a SHE(O) is a disease, disability, or untimely death that is related to an occupation and whose occurrence can provide the impetus for epidemiologic or industrial-hygiene studies or serve as a warning signal that material substitution, engineering control, personal protection, or medical care might be required (Rutstein et al., 1983). Wagener and Buffler (1989) showed how NCHS's Compressed Mortality File could be screened with SHE(O) codes to derive mortality rates for as small an area as a county, which would be a step toward identifying geographic areas with increased incidences of sentinel diseases. The industrial or occupational information routinely entered on death certificates is encoded on higher-level, more easily retrievable records in only a few states, however, and that diminishes the usefulness of the SHE(O) idea and of other occupational-surveillance programs that use mortality data. Recently, the SHE(O) principle has been adopted as the core of demonstration reporting projects between NIOSH and 10 state health departments (Baker, 1989). Lead poisoning is one of the eight target occupational-health conditions focused on by this project, the Sentinel Event Notification System for Occupational Risk (SENSOR). This trial surveillance project was described in conjunction with the reporting of a case of adult lead poisoning (blood lead was 170 µg/dl, and reporting is required for concentrations above 25 µg/dl). The index case led to finding secondary cases in the subject's workplace and family (Johnson et al., 1989).

The investigators at NIOSH and Harvard Medical School (Rutstein et al., 1983), who developed the original list of SHE(O)s, intended that new associations between toxic environmental exposures and disease conditions be added as they were recognized. The extent to which neurotoxic health effects are indistinct entities will interfere with their conversion into SHEs or SHE(O)s. Neurotoxic conditions that might be added to the list include occupational neuropathy due to Lucel 7 (Horan et al., 1985), paralysis of the urinary bladder after exposure to the NIAX catalyst dimethylaminopropionitrile (Gad et al., 1979; Pestronk et al., 1979), and the Kepone (chlordecone) syndrome (Guzelian, 1982).

• *Biologic markers.* The complexity and inaccessibility of many parts of the nervous system make it difficult to use biologic markers to study neurotoxicology. But events in physiologic systems that are substantially controlled by neuronal processes, such as some aspects of endocrine and immune function, can be monitored. And objective, computer-based systems for assessment of the function of the central and peripheral nervous systems have been developed to yield biologic markers of neurologic function in exposed populations (Letz and Baker, 1986); the tests are rapid, inexpensive, and noninvasive. Some indicators of exposure to neurotoxicants are measurable in easily sampled media, such as blood, urine, and hair. Surveillance for exposure to neurotoxicants is possible with such easily sampled surrogates, as in the measurement of peripheral esterase in workers exposed to organophosphates to monitor CNS effects indirectly (see also Chapter 3).

• *Disease and exposure registries.* Groups of people with known or suspected large exposures to neurotoxicants, such as occupationally exposed populations or residents near hazardous-waste sites, can be included in programs of targeted medical surveillance or followup, e.g., by the Agency for Toxic Substances and Disease Registry (ATSDR).

ATSDR is charged under the Comprehensive Environmental Response, Compensation, and Liability Act (CERCLA, or Superfund) with maintaining national registries of people exposed to hazardous substances. ATSDR's current registries are of people exposed to trichloroethylene, dioxin, or β-naphthylamine. The latter two are not regarded as a neurotoxicants, but the program will be expanded and could provide some unanticipated information. Available information from the NOHS (1972-1974) or the more recent NOES (1981-1983), OSHA compliance monitoring, health-hazard evaluations, and similar programs can be reviewed to identify occupational and industrial groups with the potential for substantial exposure. Another group of persons for potential followup consists of patients (probably of HMOs) who are on long-term pharmaceutical treatment regimens with potentially neurotoxic drugs. Depending on the nature of the chemical, targeted medical surveillance might consist of periodic determination of blood concentrations, measurement of cholinesterase, nerve-conduction testing, or other testing of neurologic or psychiatric function.

Prospective surveillance of populations exposed to neurotoxicants in defined settings, such as the workplace, is a promising approach to the early and sensitive detection of functional impairment of the nervous system. The methodologic advantage of prospective surveillance over one-time or cross-sectional surveillance is that each subject can serve as his or her own control. Initial information on a person can be taken to represent baseline or pre-exposure measures of function, and later data can be compared with the baseline data to detect possible toxic injury. In this study design, effects of modifiers, such as drug or alcohol abuse, on the nervous system would be largely canceled out. Thus, compared with cross-sectional evaluations, prospective surveillance should be able to detect neurologic damage at an early

stage and at low levels of exposure to such neurotoxic agents as lead in children and solvents in adults. Such surveillance is ideally targeted to the identification of functional impairment while it is still potentially reversible; detection of subclinical impairment can, in such instances, trigger a person's removal from exposure or steps to reduce overall exposures so others will not be harmed.

In another epidemiologic approach to the detection of environmental causes of neurologic disease, time trends or geographic patterns of occurrence of illness are measured. For example, in studies of multiple sclerosis, a clear north-south gradient in incidence is noted, with higher rates at more northerly latitudes. Moreover, people who spend their early years in northern latitudes and then move south seem to carry much of their earlier risk of multiple sclerosis into later life. Those observations suggest some undefined environmental factors in the etiology of this chronic neurologic disease. Likewise, strong time trends have been seen in the incidence of Parkinson's disease and motor neuron disease. The explanation is not known, but again an environmental factor is suggested (Lilienfeld et al., 1989).

Case-control studies of groups with chronic neurologic disease provide another approach to the detection of environmental causation. People with and without a chronic neurologic disease are interviewed to determine whether there are systematic and statistically significant differences in exposure between the two groups (Hertzman et al., 1990). Such studies help to detect diseases of long latency caused by exposure many years in the past.

• *Imaging techniques*. New methods are becoming available for studying the nervous system noninvasively. Computed axial tomography (CAT) allows visualization of the structure of internal organs; CAT scanning has been used, for instance, to reveal structural abnormalities in the brains of schizophrenics (Zec and Weinberger, 1986). The recently developed positron- emission tomography (PET) and magnetic resonance imaging (MRI) allow visualization of physiologic and biochemical processes as they oc- cur in the brain (Battistin and Gerstenbrand, 1986). PET scanning has been used in diagnosis of neurotoxicity in people exposed to MPTP (Calne et al., 1985). At present, however, PET and MRI are not routine screening tools.

6

Risk Assessment

Risk assessment, as defined by Crouch and Wilson (1987), is "a way of examining risks, so that they might be better avoided, reduced, or otherwise managed." It is a process based on science whose objective is to provide criteria for policy-making. Risk assessment can be qualitative or quantitative. The development of quantitative methods for risk assessment has been most extensive for carcinogenicity, that is, estimating the potential risks of cancer associated with exposure to oncogenic chemicals and other agents. *Risk Assessment in the Federal Government* (NRC, 1983) and *Identifying and Regulating Carcinogens* (OTA, 1987) discussed the development of those methods. Compared with the work on risk assessment of neurotoxicants, the development of methods for quantitative risk assessment of carcinogens has been built on a larger body of mechanistic information and hypotheses that permits quantitative description of the cellular and molecular events involved in chemical- and radiation-induced carcinogenesis. But uncertainties still attend almost all risk assessments that involve extrapolation from acute to chronic responses, from response at high doses to response at low doses, from one route of administration to another, and

from species (laboratory animals) to species (human).

In neurotoxicology, critical mechanistic information is not available, except for a few substances. Thus, risk assessment in neurotoxicology is still relatively undeveloped. Moreover, neurotoxicologic risk assessment is likely to be highly complex, compared with carcinogenicity risk assessment. Carcinogenesis at the molecular level has been proposed to involve a relatively limited set of cellular events common to the induction of all tumors; although specific carcinogens can act at different stages of carcinogenesis, the process is assumed to be fundamentally similar among tumor types and sites. In contrast, the diversity of cells and processes in the nervous system that are potentially subject to toxic action is great. Moreover, the response of those targets can depend on the developmental stage (including aging) of the organism at the time of exposure (Weiss, 1990). For example, during the cytoarchitectural organization of the human CNS after birth, neurotransmitters can function as trophic factors guiding the migration and localization of cells in brain regions; after synaptogenesis, they function primarily as information transducers (Schwartz, 1985).

An agent (such as α-methyl-p-tyrosine) that interferes with neurotransmitter synthesis during development can cause permanent alterations in regional neuronal organization; the same agent acting on the same mechanisms later in development can interrupt cell-cell communication without causing structural disarray.

No general mathematical model is likely to be appropriate for the quantitative risk assessment of neurotoxicants as a class. Defining and validating risk-assessment methods for neurotoxicants will probably require the development of a *set* of methods that are restricted in application to specific end points, rather than to specific toxicants. The rest of this chapter discusses some approaches to that task.

The process of risk assessment has been divided into the following stages (NRC, 1983; Cohrssen and Covello, 1989): hazard identification, dose-response analysis, exposure assessment, and risk characterization, followed by risk management. Dividing the process into those conceptual stages can assist in allocating resources and critical analyses to specific aspects of science-based policy analysis. In hazard identification, the goal is to obtain sufficient information to determine the qualitative nature of any biologic activity associated with a specific agent. Short-term tests are useful, including even those with a high rate of potentially false-positive results, such as the Ames bacterial mutation assay used in carcinogen hazard identification. When a hazard has been qualitatively identified, the next stage is to develop quantitative information on the relationship between dose and response. The relevance of the test system to predicting outcomes in humans—such as route and timing of exposure—becomes more important. Differences between conditions of the test and those anticipated to occur in human populations might be unavoidable, so implicit and explicit assumptions as to extrapolation must then be incorporated into experimental design and data analysis. Information on

metabolic differences between species can also be relevant. An important outcome of the analysis is the development of sufficient information to evaluate the overall shape of the dose-response relationship (linear, sublinear, hyperbolic, or sigmoidal), which can assist in generating inferences on likely responses outside the range of measured doses and responses. It should be noted, however, that the shape of the dose-response curve at the low end is often difficult to determine. In the third stage, exposure-assessment data are accumulated on external doses expected to be encountered by humans, including groups whose exposure or response might be expected to be greater than average (for instance, children). In some cases, information on internal dose (and toxicokinetics) can also be collected. Finally, all the preceding information—the nature of the expected outcome, the relationship between dose and outcome (response), and the expected ranges and distribution of doses from individual to individual—are incorporated into a risk characterization.

In the case of dichotomous events, such as the presence or absence of a diagnosable malignancy or death, the process results in a probability estimate that can be expressed as an increase in individual risk or, with appropriate adjustments, as an estimate of increased population risk. An approach based on a yes-no definition of outcome does not provide an accurate quantitation of events that vary in severity. In neurotoxicity, we are almost always concerned about outcomes other than death and effects that vary in severity (unless they are more or less arbitrarily defined in dichotomous terms, such as defining mental retardation solely on the basis of a critical IQ score). Because many neurobiologic functions are expressed on a continuum, it is likely that the effects of many neurotoxicants similarly fall on a continuum of severity. In modeling terms, neurotoxicants act to change the distribution of biologic properties that themselves assume a spectrum of values in a population. Gaylor and Slikker (1990) have provided an example of a procedure that uses data on neurochemical, neurohistologic, and behavioral effects of exposure to methylenedioxymethamphetamine (in rats or monkeys) to estimate risk as a function of dose of a potentially neurotoxic substance.

One of the most fully examined instances of this type of neurotoxicity is related to the effects of lead on cognitive function. Most measures of cognition are normally distributed around a population mean (e.g., the Stanford-Binet IQ score has a mean of 100 and a standard deviation of 15). The IQ-lowering effect of lead in children can be detected as a change in mean score (Needleman and Gatsonis, 1990). More important, however, the change in mean implies that the entire *distribution* of IQ scores in the exposed population is shifted. Although the change in *mean* IQ in lead-exposed children might seem relatively small (less than 10 IQ points in most studies and so well within the population's normal variability), an examination of the effects of such a decrease on an overall population distribution of IQ scores (Figure 6-1) reveals the potential for very important consequences of low-level lead exposure for society as a whole. The displacement of the overall curve toward lower IQ scores would reduce the proportion of children with superior IQs (i.e., 130 or more) and increase the number of children with seriously compromised intellectual potential (i.e., IQs less than 70).

Current risk-assessment methods based on carcinogenicity risk-assessment models do not capture such types of impacts, built as they are on an assumption that the end point of concern is an all-or-none (dichotomous) variable, rather than a continuous one. The problems of risk management, although not among the specific concerns of this volume, are compounded by the phenomenon of the continuum of both frequency and intensity of response related to dose. At some doses, the "most sensitive" response might be some

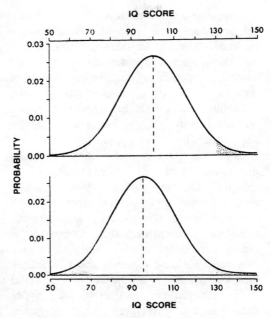

FIGURE 6-1 Effect of a shift in mean IQ score on the population distribution. The top figure represents a theoretical population distribution of IQ scores with a mean of 100 and a standard deviation of 15. In a population of 100 million, 2.3 million (stippled area) will score above 130. The bottom figure represents distribution of intelligence-test scores with a shift of 5%, yielding a mean of 95. Here, the number of individuals scoring above 130 falls to 990 thousand, with a corresponding inflation of those scoring below 70. Source: Weiss (1990).

mild form of neurotoxicity; at another, usually higher, dose, the response might be more serious, more life-threatening, or more intense.

APPROACHES TO RISK ASSESSMENT FOR NEUROTOXICITY

Risk-assessment models ideally are based on two sources of biologically based information: statistical analyses of rich data sets and knowledge of mechanisms. Approaches involving curve-fitting require a great deal of data to develop predictions of risk with any reliability. Although, in appropriate cases, statistical analysis can be important in validating models based on hypothesized mechanisms of action, the available data are usually insufficient to resolve critical issues, such as the predicted response at exposures below those on which the data were collected. An essential part of developing risk-assessment methods is therefore the integration of biologically based models that incorporate our understanding of mechanism of action. It is important to emphasize that statistical models should conform to biologic hypotheses—that is, curve-fitting must be consistent with what is known of the biology of the affected system.

The most commonly used approach for risk assessment of neurotoxicants, as of most noncarcinogens, is the uncertainty- or safety-factor approach (NRC, 1986; Kimmel, 1990). It is based on identifying either the lowest dose of a toxicant at which adverse effects are observed (lowest observed-adverse-effect level, LOAEL) or the highest dose at which no adverse effects are observed (no-observed-adverse-effect level, NOAEL) and then deriving a presumably safe dose by dividing by a safety or uncertainty factor. Its weakness is related to the unreliability of observation as a means of determining doses at which biologic effects actually do or do not occur. A small study might give negative results, and no study is large enough to exclude the possibility of any effect. Furthermore, the importance of subtle alterations in behavior or small neurochemical or structural changes is a matter of debate. Before undertaking tests to determine the LOAEL or NOAEL, it is important to

define what is meant by "adverse effect" and to develop appropriately sensitive measures of such effects. For example, the persistence of the neurotoxic effects of small exposures to such solvents as trichloroethylene, dichloromethane, and toluene is not known, so the importance of observed decrements in performance is still under debate (Gade et al., 1988; Parkinson et al., 1990). This approach also provides no information on the slope of the dose-response curve or intensity of effect above the "safe" dose calculated (Gaylor and Slikker, 1990; Kimmel, 1990; IOM, 1991).

Thus, the determination of a LOAEL or NOAEL is strongly influenced by the sample size and design of the experiment (Gaylor, 1983; Crump, 1984). In most cases, it is difficult to determine an effect smaller than a 20% increase or decrease. As an alternative, Crump (1984) first suggested the use of a benchmark dose (BD), defined as "a statistical lower confidence limit corresponding to a small increase in effect over the background level." The increase in effect used to determine the BD would be near the lower limit of change, which can be determined with reasonable accuracy in toxicologic studies. The BD is calculated with a mathematical model; however, because BDs correspond to risks in the experimental range, they are less affected by the particular shape of the dose-response model used in the calculations. The acceptable dose is extrapolated from the BD. Using a BD, rather than the more usual LOAEL or NOAEL, avoids problems inherent in defining a negative (the NOAEL) and in the lack of precision associated with defining a LOAEL (Kimmel, 1990).

Whether a conventional approach or a BD approach is used, the dose identified is usually modified by some safety factor to yield a so-called safe dose. The incorporation of a safety factor (usually 10 raised to some power) reflects a policy decision to incorporate in a quantitative manner some of the uncertainties associated with risk assessment. In general, safety factors range between 10 and 1,000 (NRC, 1977; Kimmel, 1990), so the risk assessment is the final estimate of dividing the BD, NOAEL, or LOAEL by 10, 100, or 1,000. Selection of the safety factor involves processes of judgment, including considerations of weight of evidence, availability of information on human response, type of toxicity observed, and probability of variations in response among susceptible groups. Some rules of thumb have been adopted to accommodate intraspecies variability, cross-species extrapolation, experimental duration, etc.; they correspond fairly well to observed variation (Dourson and Stara, 1983).

As noted above, statistical approaches to risk assessment should be consistent with what is known about mechanisms of action or about the biology of the affected system. The NOAEL-LOAEL approach is based on an assumption of a threshold, a dose below which an effect does not change in incidence or severity. However, the evidence of the general applicability of that assumption for all neurotoxicants is relatively weak. Even though a given neurotoxic response may require a threshold dose of a specific toxicant, other toxicants in the environment that cause the same or similar response may in effect lower the threshold dose of the specific neurotoxicant of interest. That is, a person may have a threshold of effect in a pure environment, but there may not be a threshold for a heterogeneous population in a heterogeneous environment. It is likely that some neurotoxicants have thresholds and others do not. (A lack of a threshold for a neurotoxicant makes the safety-factor approach biologically indefensible.) On the one hand, the nervous system appears to have considerable reserve capacity and plasticity, which would support the assumption of a threshold. On the other hand, it is composed of cells that are nonreplaceable, which would argue that no damage can be considered innocuous. Moreover, both susceptibility to injury and the nature of the injury can vary with developmental stage;

early neurologic development includes a selective pruning of apparently excess neurons and connections (Rakic and Riley, 1983), whereas senescence includes a progressive loss of specific neurons and an apparently selective loss of function. Because the CNS develops from a limited set of cells early in development, it is appropriate to assume that exposure to neurotoxicants during critical periods of brain development can be without a threshold; data on x-irradiation-induced brain injury support this assumption (Schull et al., 1990). As shown in Figure 6-2, when in utero radiation exposure occurred during the period between 8 and 15 weeks gestation, the effects on intellectual status

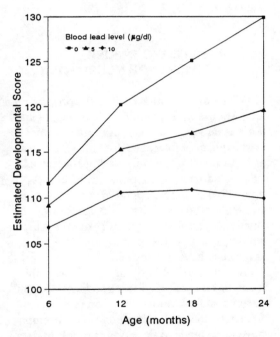

FIGURE 6-2 Estimated developmental scores at various ages for three blood-lead concentrations. Application of a model developed by Hattis and Shapiro (1990) to data from Bellinger et al. (1987a,b) that assumes the group is exposed to yield constant blood lead concentrations of 0, 5, and 10 μg/dL leads to estimates of developmental scores for populations of exposed children. Source: Adapted from Wyzga (1990).

were shifted to the left, possibly without a threshold. Whether the PNS and CNS differ with regard to thresholds is not known. Studies of PNS neurotoxicants, such as acrylamide, suggest a capacity for recovery through axonal regeneration, but it varies with the dose and duration of exposure; Hattis and Shapiro (1990) have investigated quantitative dose-response models as alternatives to the NOAEL-LOAEL and uncertainty-factor approach for this situation. The recent Institute of Medicine (IOM, 1991) publication on safety of seafood takes a similar quantitative modeling approach for methylmercury. In any event, the application of the NOAEL-LOAEL (or BD) approach to all neurotoxicants is unlikely to be biologically defensible.

CURVE-FITTING IN RISK ASSESSMENT FOR NEUROTOXICITY

When quantal information (the proportion of response at a given dose) is available, but little is understood about underlying biologic mechanisms, a tolerance distribution, such as the probit (log-normal) or logistic models, can be adopted to fit population data (Wyzga, 1990). Those sigmoid dose-response models assume that the distribution of individual thresholds follows the specified model, but that the population as a whole might not have a threshold.

If a substantial body of data is available, it might be possible to test goodness-of-fit of various statistical models. For most neurotoxicants, that is not now possible. Attempts have been made to fit various models to data on the PNS and CNS toxicity of lead. Data on nerve-conduction velocity in children exposed to environmental lead (through living near a smelter) as a function of blood lead concentration fit a "hockey-stick" type of dose-response curve (Schwartz et al., 1988). The data were sufficient to permit discrimination, in terms of goodness-of-fit, from several other models, including a logistic

model and a quadratic model. In another approach to assessing lead neurotoxicity, Wyzga (1990) assumed that individual thresholds for a dichotomized measure, muscle weakness, were log-normally distributed in the population. After producing maximal-likelihood estimates of the mean and standard deviation of the threshold distribution, the distribution of blood lead concentrations was estimated. Wyzga thereby determined that the probability that a randomly chosen male from the general population would exceed the threshold and exhibit extensor muscle weakness would be 0.01. In a second analysis, Wyzga applied the model developed by Hattis and Shapiro (1990) for acrylamide exposure that compensates for past damage with the rate of repair proportional to the cumulative-damage data on children's blood lead concentrations and mental development gathered by Bellinger et al. (1987a,b). The dose-response estimation related group response to group exposure levels, and Wyzga concluded that this model, albeit only one of many alternative models that could be fitted to the data, indicates some inhibition of development at exposures that resulted in very low blood lead concentrations.

The hockey-stick model was also evaluated for its consonance with the available information on lead's effects and on peripheral neurophysiology. As discussed by Schwartz et al. (1988), it is plausible to propose a threshold for the peripheral neurotoxicity of lead for reasons of both mechanism of action and physiology: lead inhibits synaptic release of acetylcholine through inhibition of ion-dependent neurotransmitter release, and lead reduces velocity of peripheral-nerve conduction (as measured transdermally with surface electrodes). The mechanism of action of lead thus involves events that are nonlinear: the changes in ion permeability that are associated with nerve depolarization and transmitter release involve many discrete events (e.g., channel openings, exocytotic events of electric propagation), and nerve conduction is a summation involving the function of many nerve fibers. Thus, both basic biology and mechanism of action are consistent with the results of curve-fitting in this case.

MECHANISTIC MODELS FOR RISK ASSESSMENT FOR NEUROTOXICITY

Information on mechanisms of action provides the most powerful tools for developing models for risk assessment (Silbergeld, 1990). For neurotoxicants, it is unlikely that a single mechanism or even a small number of mechanisms will be elucidated for the neurotoxicants already identified, given the variety of potential cellular and molecular targets of action. Nevertheless, in some cases, information is available on mechanisms of action of some neurotoxicants and can be used for purposes of developing quantitative approaches to risk assessment. It must be emphasized that the generalizability of such models is narrow, possibly even for similar molecules, so the approach might be useful only for assessing, case by case, the risks of the substances for which it was developed. Further developments in structure-activity analysis might permit extension of specific mechanistic models to chemicals within appropriate structural classes.

As noted in Chapter 1, many pesticides are designed to be neurotoxic to a target organism. They should be relatively nontoxic to humans in their intended uses, but will be neurotoxic to humans who are sufficiently exposed. The reason is that their mechanisms of action in target species involve biologic substrates that are also present in humans.

For example, the organochlorine pesticide dieldrin once was widely used in agriculture and in household formulations, but now is restricted in application because of its persistence and potential carcinogenicity. Dieldrin acts as a neurotoxicant by blocking the GABA channel in nerve cells (Narahashi and Frey, 1989). It is a competitive antagonist

for the recognition site (receptor) associated with the channel, and it can be shown to bind to the site with a high affinity. Assuming that to be its fundamental mechanism of action, one can develop a mathematical model based on the biology of receptor-ligand interactions. Receptor-ligand binding is well described by classical mathematical models derived from Lineweaver-Burk equations of enzyme kinetics (Silbergeld, 1990). (Enzyme-substrate interactions are a class of receptor-ligand interactions, as is carcinogen-adduct binding to DNA.) As with all receptor-ligand interactions, when the concentration of the receptor is greater than that of the ligand at low concentrations of dieldrin, receptor-ligand binding is linearly related to the concentration of the ligand, dieldrin. If receptor binding is directly related to regulation of the ionophore and the function of the ionophore is critical to neuronal function, a linear model based on receptor-binding kinetics can be constructed for the neurotoxic risks associated with dieldrin. The model agrees well with electrophysiologic data. This approach provides a prediction of risk that can be validated in terms of whole-organism response to dieldrin.

It might be possible to model other neurotoxicants that act through receptor mechanisms with similar approaches to quantitative risk assessment. (Table 6-1 lists such neurotoxicants.) Obviously, these approaches describe the predicted behavior of a toxicant only at its cellular site of toxic action, e.g., relations between target-organ dose and response. For risk assessment, knowledge of pharmacokinetics—absorption, distribution, and metabolism—is relevant for estimating the overall relationship between external exposure and target-organ dose (NRC, 1987b).

Another mechanism of neurotoxic action could be applicable to an understanding of the consequences of cytotoxic agents that act on neural cells during development. On the basis of the principle of nonreplaceability, a

TABLE 6-1   Some Neurotoxicants That Act on Receptors

Receptor or Channel	Blocker (Antagonist)	Modulator (Agonist)
Acetycholine receptor	Lacticotoxin Erabutoxin α-Conotoxins Anatoxin-a Nereistoxin Atropine Scopolamine	
Acetycholine-activated channel	Histrionicotoxin Amantidine N-Alkylguanidines	
Excitatory amino acid	AP-5 (2-amino-5- phosponopentanoate) AP-7 (2-amino-5- phosponopentanoate) Nephila orb web spider toxins Argiope orb web spider toxins γ-Philanthotoxin MK 801 Ketamine	Oxotremorine
GABA$_A$ receptor or channel	Bicuculline Lindane Dieldrin Picrotoxinin	N-Methyl-D-aspartate 1-Glutamate Kainate Quisqualate
GABA$_B$ receptor or channel	Phaclofen	Muscimol Avermectin B^{1a} Barbituates Benzodiazepines Ethanol
Glycine receptor or channel	Strychnine	Baclofen
Presynaptic tunnel	β-Bungarotoxin Botulinum toxin Tetanus toxin Taipoxin	α-Latrotoxin

linear no-threshold model might be appropriate to estimate the effects of cytotoxic agents encountered during critical periods of development. Radiation is the classic example of such an exposure. The results of animal studies are consistent with the hypothesis of a lack of threshold for prenatal irradiation during the critical period of organogenesis of the cortex (Schull et al., 1990). The available human data are also consistent with the hypothesis (Otake and Schull, 1984; NRC, 1990). As shown in Figure 6-3, there is no clear evidence of a threshold for the effects of prenatal irradiation encountered during the critical period of fetal development, with respect to mental retardation, in children irradiated in utero as a consequence of the atomic bombs in Hiroshima and Nagasaki.

Of greatest interest and challenge in the development of appropriate models is the possibility, discussed in Chapter 2 of this report, that early neurotoxic exposure can result in cumulative or progressive damage expressed as a late-stage degenerative disease, such as dementia or a major motor disorder. Clinical data on the parkinsonism-dementia complex on Guam and on MPTP intoxication in drug abusers in the United States indicate that such damage does occur. Although models for cumulative damage have not been extensively developed for neurotoxicity, Hattis and Shapiro (1990) exploited the abundance of existing data on acrylamide-induced neurotoxicity in developing a model that is relevant for reversible, cumulative neurotoxic effects. Alternatively, some of the time-to-tumor models (Peto et al., 1980; Krewski et al., 1983) considered in

cancer risk assessment might be relevant to such mechanisms of neurotoxicity.

Design and reporting of both animal and epidemiologic experiments could be modified in several respects to facilitate modeling of neurotoxic end points for risk assessment (Wyzga, 1990). It would be useful to conduct studies at a number of carefully measured exposure concentrations, including ones of environmental relevance. Recording of specific quantitative responses is more informative than dichotomizing or categorizing observations, and making data available in nonaggregated form would enable thorough exploration and interpretation of research data by other researchers. Summary statistics are sufficient only if the model assumed to condense the data is appropriate, and that is seldom known with certainty. Comparable sets of animal and human data should be analyzed in parallel, to establish the most effective methods of using animal data when human data are not available.

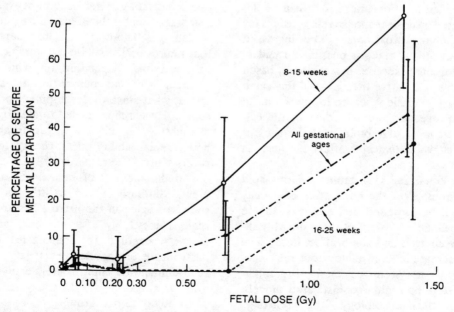

FIGURE 6-3 Percentage of severe mental retardation among those exposed in utero by dose and gestational age in Hiroshima and Nagasaki. Vertical lines indicate 90% confidence intervals. Source: NRC (1990).

SUMMARY

The critical focus for basic and applied research in neurotoxicology remains in the primary stage of risk assessment, that is, hazard identification. Few chemicals have been adequately tested for potential neurotoxicity, and data even on many of those are still insufficient for evaluating risk. In basic research, we still need investigations to develop models for integrating molecular, biochemical, cellular, and organ-level events into rational explanations of behavior, function, and learning. Applied research is needed in both clinical and experimental disciplines, to develop appropriate methods for the detection of neurotoxicity, particularly in developing and aging organisms. Efforts in the risk assessment of neurotoxicants must focus first on improving hazard identification. Concerted efforts at improving and validating tests and increasing the vigilance and sensitivity of epidemiologic surveillance must be undertaken. With improved hazard identification, more data can be accumulated on dose-response relationships for particular neurotoxicants. The increased data base might make it possible to undertake additional exercises in statistically based modeling and curve-fitting. At the same time, further basic research in neuroscience will improve our understanding of the cellular and molecular biology of the sites of action at which toxicants affect the nervous system.

The visual and visuomotor systems hold great promise for the integration of toxicity data and fundamental neurobiology. Some chemicals affect vision, others perturb visuomotor control, and some affect both. The great increase in knowledge about the mechanics of these systems has enabled researchers to develop highly sophisticated models, including the neurobiology of the retina (e.g., Stryer, 1983), the organization of the visual cortex (e.g., Hubel and Livingstone, 1987), and the neuromotor control of visual tracking (e.g., Fromm and Evarts, 1981). At least

two neurotoxicants—methylmercury (Friberg, 1977) and methanol (Politis et al., 1980)—affect visual function. However, no attempt has yet been made to test the fit of existing toxicity data on those two agents to any of the complex neurobiologic models to predict the results of small exposures or exposures at different stages of development.

Mechanistic understanding is the most powerful tool in developing models for quantitative risk assessment, for evaluating predictions of dose-response relationship outside the measurable range, and for extrapolating across species reliably. Model-building in science is important in identifying critical data gaps. Neurotoxicants have always served an important role as tools in basic neurobiology (Narahashi, 1989); and basic neurobiology is essential in neurotoxicity risk assessment. Attempting risk assessment for neurotoxicants can assist in the identification of critical data gaps to be addressed in developing and validating mechanistic models of neurotoxicity. For instance, one important general issue in neurotoxicity risk assessment is the potential irreversibility of damage. There are reasons to anticipate that neurotoxicity will be irreversible or will be reversible, depending on which components of the nervous system are affected and at what stage of development or senescence. Research is needed to identify which elements of the CNS are irreplaceable and how irreplaceability might change to affect reversibility. The research will also increase our understanding of neurologic development. Similarly, we know little about processes of aging in the brain. If some brain regions inevitably lose neurons, are they more susceptible to neurotoxic damage, particularly of the delayed type? Would such damage be appropriately estimated by a linear approach based on the assumption of additivity to background?

The greatest need in developing quantitative approaches to risk assessment of neurotoxicants is for data on neurotoxic agents. The intensive study of well-characterized neurotoxicants can be of great importance, as demonstrated by recent research on lead and *n*-hexane. In studies of neurotoxicants, attention should be given to the concentration, duration, and timing of exposure. Especially rich data sets on specific neurotoxicants should be examined for opportunities to test goodness-of-fit (statistical) models, as has been done for lead.

Efforts should be directed to developing a broader range of biologically based models for risk assessment of neurotoxicants, with emphasis on nondichotomous events. It might be useful to examine the research on model-building that is already under way in neuroscience, particularly cognitive science. Caution should be exercised in relying on simple uncertainty-factor and threshold models for the risk assessment of all neuro-toxicants under all exposure conditions in all populations.

Basic neurobiologic research (e.g., research on the neurobiology of cortical and cerebellar development and on the neuronal changes in the aging brain) should include investigation of the effects of toxicants. Research should be undertaken to further understanding of the role of pharmaco-kinetics in neurotoxicity so as to encourage the development of physiologically based pharmacokinetic modes such as are being developed to aid in carcinogenicity risk assessment, as well as of the differences and similarities between animals and humans with regard to neurotoxicity. Research should also be undertaken to test the utility of cellular models, such as receptor-binding models, for hazard identification and for predicting an organism's overall response.

7

Conclusions and Recommendations

The Committee on Neurotoxicology and Models for Assessing Risk identified the following six major gaps in information or conceptual development in environmental neurotoxicology (listed in a sequence to parallel the structure of the report):

• The extent to which disease and dysfunction of the human nervous system are the result of exposure to toxic environmental agents
• The neurotoxicity of most chemicals in commerce
• Biologic markers of neurotoxic exposures or effects that could be applied in epidemiologic and clinical studies
• Well-designed, hypothesis-driven neuroepidemiologic research to increase knowledge about the scope of the problem of neurotoxicity
• Strategies for the rational, efficient testing of chemical substances for neurotoxicity
• Risk-assessment paradigms for evaluation of neurotoxic end points.

The committee's specific conclusions and recommendations to address these gaps are as follows.

The committee concludes that neurotoxic effects can be caused by exposure to chemical agents in the environment. Environmental chemicals have been shown to cause neurotoxic effects in individual cases and in epidemics. Neurotoxicity caused by environmental toxicants results in a range of neurologic and psychiatric disorders; the complexity of the disorders reflects the enormous diversity of the nervous system's functions and the presence in the nervous system of a large number of cellular and subcellular targets. Neurotoxic outcomes range from devastating illnesses, such as parkinsonism and dementia, to subtle changes, such as alterations in behavior and limitations on memory and cognition. In addition to immediate and progressively developing effects, there is increasing evidence that neurotoxic effects can occur after long latent periods. It is postulated that intervals as long as many decades can elapse between exposure to a chemical and the appearance of neurologic illness. Concern over the potential neurotoxic effects of chemical substances is greatest for agents that cause irreversible or progressive changes. Chemicals can permanently alter brain development and cause subclinical dysfunction, or they can reduce reserve capacity of the nervous system, which

123

may become manifest as disease in the elderly. On the basis of the available evidence, the committee hypothesizes that an as-yet-unspecified fraction of human neurologic and psychiatric disease is attributable to chemical agents in the environment.

The committee concludes that a major obstacle to assessing the extent to which chemicals in the environment cause nervous system diseases and dysfunction is that little qualitative or quantitative information is available on possible adverse effects of most environmental chemicals on the nervous system. Some chemicals in commerce are known to have neurotoxic potential, but most commercial chemicals have not been assessed for neurotoxicity. Although resources are not readily available to undertake across-the-board testing of all chemicals already in commerce, prudent public policy dictates that all chemicals, both old and new, be subject to at least basic screening for neurotoxicity when use and exposure warrant. There is a particular lack of data on chronic and long-latency neurotoxic effects. Structure-activity relationships, now the most widely used approach to assessment of toxicity, provide a poor basis for predicting neurotoxic potential; however, greater understanding of chemical structural correlations and the underlying mechanisms of toxicity can be expected to lead to the discovery of more useful applications of structure-activity relationships.

The committee recommends that a more accurate estimate of the extent of the problem of neurologic and psychiatric dysfunction attributable to chemical agents in the environment be made.

The estimate must be based on a combination of clinical, epidemiologic, and toxicologic studies coupled with the techniques of quantitative risk assessment.

The committee concludes that additional biologic markers for the assessment of subclinical neurotoxic effects are needed. Such markers can be biochemical, structural, or functional. They can be developed either through in vitro analyses, through animal studies, or during observational studies in human populations exposed to environmental neurotoxicants. Although associations between biologic markers and disease are usually established initially in cross-sectional studies, a particular need exists to validate putative biologic markers in prospective studies. Only in longitudinal prospective studies can the ability of biologic markers to predict the occurrence of disease be accurately assessed.

The committee recommends that putative biologic markers in animal species be evaluated and validated in in vivo and in vitro systems. The committee further recommends that biologic markers be regularly incorporated into epidemiologic and clinical studies of neurologic disease, particularly prospective studies.

The primary goal of the incorporation of biologic markers into such studies should be to validate their predictive accuracy and to test hypothesized quantitative relationships between specific markers that are likely to be on or close to causal pathways and neurotoxic outcomes.

The legislated authorities of the federal regulatory agencies (the Occupational Safety and Health Administration, the Environmental Protection Agency, the Consumer Product Safety Commission, and the Food and Drug Administration) to regulate or to require neurotoxicity testing might be too narrow to encourage investigation of neurotoxicity. The potential authority of the existing mandates of those agencies to control toxic substances, including neurotoxicants, has not been fully used.

The committee concludes that tests are available to construct a tiered approach to neurotoxicity testing. The first tier, or screen, is intended for hazard identification. The results of the screen and a chemical's exposure pattern would determine further characterization of dose-response relationship (second tier) and mechanisms (third tier). No validated system satisfies all the necessary requirements for a screening program to detect the neurotoxic potential of chemicals. The range of such a program should extend to the detection of neurodevelopmental effects and effects on cognitive function and of neuroendocrine effects. No comprehensive effort has yet been made to determine the predictive ability of individual screening tests by examining the relationship between test results and data from long-term studies in animals or epidemiologic and clinical studies in humans.

The committee recommends that a rational, cost-effective neurotoxicity testing strategy be developed and adopted.

It should allow an accurate and efficient progression from the results of hazard-identification studies (screening) to the selection and application of appropriate test methods for defining mechanisms of toxicity and for quantitative characterization of neurotoxic hazards.

Benefits to be gained are several. First, it would permit development of tests that would confirm or disprove each other at each of several levels of complexity; thus, it would incorporate a series of checks and balances. Second, the results of the later phases of testing would provide the data necessary to evaluate and validate initial screening batteries and thereby help to identify tests that should be excluded from or incorporated into an efficient battery. Third, the information generated by such a strategy would reveal which types of data are most useful for accurate, quantitative prediction of the risks to humans associated with exposure to similar chemical compounds.

For reasons of efficiency, integrative studies combining a variety of end points need to be explored in the development of the strategy. For example, tests of neurotox-

icity on chronically exposed animals might be carried out in conjunction with tests of other chronic effects. An objective testing strategy rationally applied would enhance risk assessment by providing data rapidly for decision-making. Moreover, testing would identify toxic effects of substances not previously known to be neurotoxic. To maximize detection of toxicity, some toxicity studies encompassing the full life span of experimental animals should be encouraged.

The committee recommends that existing in vitro test methods be exploited more extensively than at present to identify and analyze the mechanisms of neurotoxic action at cellular levels.

To that end, it is necessary to undertake a program to test the relationship between in vitro and in vivo findings and between animal and human results for a set of well-defined substances. It should be understood that some in vitro test methods will be most useful only for screening substances and will have little application in assessing mechanisms. It is necessary to question how effective those methods will be in predicting chronic effects in whole animals--a continuing objective of method validation. Additional validation might be accomplished by determining the concordance between the results of tests and the findings of epidemiologic studies.

The committee recommends that studies to define mechanisms of neurotoxicity in as much detail as possible be encouraged, as well as studies to identify hazards.

Data are needed on the influence of dose, route of exposure, toxicokinetics, metabolism, and elimination on the effects of a given neurotoxicant, and data are needed on interspecies differences. Moreover, a detailed understanding of the pathogenesis of the neurotoxic injury caused by various agents is needed. What are the toxic metabolites? What are the molecular targets within the nervous system? How does the reaction of a toxicant or metabolite with a target trigger the sequence of events that lead to functional changes or degenerative events? What is the sequence of events? What biologic factors determine vulnerability, resistance, and capacity for reversibility and repair? What is the basis for interspecies differences? Such understanding will make possible the better prediction of relationships between exposure and neurotoxic response, the development of strategies to prevent exposure, and eventually the treatment of the exposed.

Development of a mechanistic understanding of neurotoxicity might facilitate the discovery of biologic markers of exposure to toxicants, as well as markers of early, subclinical neurotoxic effects. More complete understanding of neurotoxic disease at a molecular level should also improve the ability to evaluate new chemicals on the basis of structure-activity relationships, which as currently used can provide only minimal guidance for hazard identification.

The committee concludes that attempts to quantify the exposure of populations to neurotoxic chemicals have been limited. Clinical evaluation of neurotoxic illness and epidemiologic surveillance of populations at high risk for neurotoxicity have been fragmentary and inadequate. Few attempts have been made to explore the possible relationships between chemical exposures and chronic or progressive neurologic and behavioral disorders. The disorders include developmental delays in the young and some forms of dementia and parkinsonism in the elderly.

The committee recommends that exposure-surveillance systems cover a much broader range of chemicals and use improved monitoring techniques for long-term assessment.

The committee also recommends that

existing disease-surveillance systems, such as those of the Social Security Administration, the Department of Veterans Affairs, and the National Center for Health Statistics, be modified to provide more useful data on the incidence and prevalence of chronic neurologic and psychologic disorders, some of which are likely to be of occupational and environmental origin.

A broader range of neurologic disease end points should be covered by surveillance programs.

The committee further recommends that anecdotal reports of neurotoxicity in humans be pursued vigorously with clinical surveillance and followup.

The first step is to identify the substance or combination of agents apparently responsible for observed effects. When animal models of the human disease have been established, laboratory studies can then determine the mechanistic details that would assist in controlling the original situation and facilitate risk assessment.

The committee concludes that recognition of the possible environmental origin of neurologic and psychiatric disease is hampered by the inadequate training of most physicians and other health providers in occupational and environmental medicine. The committee concludes that uniform clinical definitions of neurotoxic disorders are needed, to provide a common basis for reporting by physicians.

The committee recommends that improved disease reporting be supported by the dissemination of information on neurotoxic illnesses to physicians and other health professionals, to increase their awareness of environmental neurotoxicity as a possible explanation of specific illnesses or sets of symptoms.

The committee further recommends that

all physicians should be trained to take a thorough occupational-exposure history and to be aware of other possible sources of toxic exposure, such as hobbies and self-medication.

The committee recommends that standardized national reporting systems be established for physicians to report outbreaks of suspected environmentally and occupationally caused neurologic and psychiatric disorders.

The incorporation in surveillance systems of the concept of sentinel health events (SHEs) specifically for neurotoxic illnesses should be encouraged.

The committee concludes that commonly used paradigms for risk assessment do not accurately or adequately model the risks associated with exposure to neurotoxicants. Neurotoxicologic risk assessment has been largely limited to the application of no-observed-effect levels and uncertainty factors, so it has not generated specific risks for given magnitudes of exposure.

The committee recommends that further attention be paid to experimental designs for studies of neurotoxic agents that provide information needed in the risk-assessment process.

Such variables as age, sex, duration of exposure, and route of exposure need to be more systematically evaluated. Species-specific effects need to be recognized. Ex-periments should include a range of doses that spans those relevant to expected human exposures. In addition to providing a firmer basis for estimating human risk, such designs permit tests of the assumption that results obtained at high doses predict the pattern of effects at low doses. The usefulness of that kind of information for quantitative risk assessment would be greatly amplified by serial measures of neurotoxic end points and biologic markers.

The committee recommends that researchers make complete, original data sets available to other investigators to facilitate full exploration of relationships and development of risk-assessment models.

The committee further recommends that to improve the assessment of the human risks associated with exposure to possible neurotoxic agents, risk-assessment methods that capture the complexities of the neurologic response, including dose-time-response relationships, multiple outcomes, and integrated organ systems be developed.

A single model will not be adequate for all conditions of exposure, for all end points, or for all agents. It might be necessary to build risk-assessment models to deal simultaneously with several end points produced by a toxicant. Such models should incorporate biologic markers of neurologic dysfunction and be based on fundamental information on mechanisms derived from experimental test systems and epidemiologic data.

References

Abou-Donia, M. B. 1981. Organophosphorous ester-induced delayed neurotoxicity. Annu. Rev. Pharmacol. Toxicol. 21:511—548.

ACGIH (American Conference of Governmental Industrial Hygienists). 1982. Threshold Limit Values for Chemical Substances and Physical Agents in the Workroom Environment with Intended Changes for 1982. Cincinnati, OH: American Conference of Governmental Industrial Hygienists Publications Office.

ACS (American Chemical Society), Committee on Environmental Improvement and Subcommittee on Environmental Analytical Chemistry. 1980. Guidelines for data acquisition and data quality evaluation in environmental chemistry. Anal. Chem. 52:2242—2249.

Adams, J. 1986. Methods in behavioral teratology. Pp. 67—92 in Handbook of Behavioral Teratology, E. P. Riley and C. V. Vorhees, eds. New York: Plenum.

Aghajanian, G. K. 1972. LSD and CNS transmission. Annu. Rev. Pharmacol. 12:157—168.

Aldridge, W. N. 1990. An assessment of the toxicological properties of pyrethroids and their neurotoxicity. Crit. Rev. Toxicol. 21:89—104.

Aleu, F. P., R. Katzman, and R. D. Terry. 1963. Fine structure and electrolyte analysis of cerebral edema induced by alkyltin intoxication. J. Neuropathol. Exp. Neurol. 22:403—413.

Allen, N., J. R. Mendell, J. Billmaier, R. E. Fontaine, and J. O'Neill. 1975. Toxic polyneuropathy due to methyl n-butyl ketone. Arch. Neurol. 32:209—218.

Altenkirch, H., S. Stoltenburg-Didinger, and C. Koeppel. 1988. The neurotoxicological aspects of the toxic oil syndrome (TOS) in Spain. Toxicol. 49:25—34.

Amdur, M.O., J. Doull, and C.D. Klaassen, eds. 1991. Casarett and Doull's Toxicology: The Basic Science of Poisons, 4th ed. New York: Pergamon Press. 1033 pp.

Anger, W. K. 1984. Neurobehavioral testing of chemicals: Impact on recommended standards. Neurobehav. Toxicol. Teratol. 6:147—153.

Anger, W. K. 1985. Neurobehavioral tests used in NIOSH-supported worksite studies, 1973-1983. Neurobehav. Toxicol. Teratol. 7:359—368.

Anger, W. K. 1986. Workplace exposures. Pp. 331—347 in Neurobehavioral Toxicology, Z. Annau, ed. Baltimore, Md.: Johns Hopkins University Press.

Anger, W. K. 1989. Human neurobehavioral toxicology testing: Current perspectives. Toxicol. Ind. Health 5:165—180.

Anger, W. K. 1990. Worksite behavioral research: Results, sensitive methods, test batteries, and the transition from laboratory data to human health. Neurotoxicology 4:627—718.

Anger, W. K., and B. L. Johnson. 1985. Chemicals affecting behavior. Pp. 51—148 in Neurotoxicity of Industrial and Commercial Chemicals, Vol. 1, J. L. O'Donoghue, ed. Boca Raton, Fla.: CRC Press.

Annau, Z., and C. U. Eccles. 1986. Prenatal exposure. Pp. 153—169 in Neurobehavioral Toxicology, Z. Annau, ed. Baltimore: The Johns Hopkins University Press.

Anthony, D.C., and D.G. Graham. 1991. Toxic responses of the nervous system. Pp. 407—429 in Cassarett and Doull's Toxicology: The Basic Science of Poisons, 4th ed., M.O. Amdur, J. Doull, and C.D. Klaassen, eds. New York: Pergamon Press.

Anthony, D. C., K. Boekelheide, and D. G. Graham. 1983a. The effect of 3,4-dimethyl substitution on the neurotoxicity of 2,5-hexanedione. I. Accelerated clinical neuropathy is accompanied by more proximal axonal swellings. Toxicol. Appl. Pharmacol. 71:362—371.

Anthony, D. C., K. Boekelheide, C. W. Anderson, and D. G. Graham. 1983b. The effect of 3,4-dimethyl substitution on the neurotoxicity of 2,5-hexanedione. II. Dimethyl substitution accelerates pyrrole formation and protein crosslinking. Toxicol. Appl. Pharmacol. 71:372—382.

Arezzo, J. C., and H. H. Schaumburg. 1980. The use of Optacon[z] as a screening device: A new technique for detecting sensory loss in individuals exposed to neurotoxins. J. Occup. Med. 22:461—464.

Arezzo, J. C., and H. H. Schaumburg. 1989. Screening for neurotoxic disease in humans. J. Am. Coll. Toxicol. 8:147—155.

Arezzo, J. C., H. H. Schaumburg, and C. A. Petersen. 1983. Rapid screening for peripheral neuropathy: A field study with the Optacon. Neurology 33:626—629.

ASPH (Association of Schools of Public Health). 1988. A proposed national strategy for the prevention of neurotoxic disorders. Pp. 31—48 in Proposed National Strategies for the Prevention of Leading Work-Related Diseases and Injuries, Part 2. Association of Schools of Public Health, Washington, D.C.

Baker, E. L. 1989. Sentinel event notification system for occupational risks (SENSOR): The concept. Am. J. Public Health 79(Suppl.):18—20.

Baker, E. L., R. E. Letz, A. T. Fidler, S. Shalat, D. Plantamura, and M. Lyndon. 1985. A computer-based neurobehavioral evaluation system for occupational and environmental epidemiology: Methodology and validation studies. Neurobehav. Toxicol. Teratol. 7:369—377.

Barbeau, A., M. Roy, G. Bernier, G. Campanella, and S. Paris. 1987. Ecogenetics of Parkinson's disease: Prevalence and environmental aspects in rural areas. Can. J. Neurol. Sci. 14:36—41.

Battistin, L., and F. Gerstenbrand, eds. 1986. PET and NMR: New Perspectives in Neuroimaging and in Clinical Neurochemistry. New York: Alan R. Liss. 518 pp.

Bellare, R. A. 1967. Studies in Manganese Poisoning. University of Bombay, India. Ph.D. Dissertation.

Bellinger, D., A. Leviton, C. Waternaux, H. Needleman, and M. Rabinowitz. 1987a. Longitudinal analyses of prenatal and postnatal lead exposure and early cognitive development. N. Eng. J. Med. 316:1037—1043.

Bellinger, D., J. Sloman, A. Leviton, C. Waternaux, H. Needleman. 1987b. Low-level lead exposure and child development: Assessment at age 5 of a cohort followed from birth. Proceedings of the International Conference on Heavy Metals in the Environment, Vol. 1. Edinburgh: CEP Consultants. (Abstr.)

Belongia E. A., C. W. Hedberg, G. J. Gleich, K. E. White, A. N. Mayeno, D. A. Loegering, S. L. Dunnette, P. L. Pirie, K. L. MacDonald, and M. T. Osterholm. 1990. An investigation of the cause of the eosinophilia-myalgia syndrome associated with tryptophan use. N. Engl. J. Med. 323(6):357–365.

Betz, A. L., G. W. Goldstein, and R. Katzman. 1989. Blood-brain-cerebrospinal fluid barriers. Pp. 591–608 in Basic Neurochemistry, 4th ed., G. Siegel, B. Agranoff, R. W. Albers, and P. Molinoff, eds. New York: Raven Press.

Bhattacharyya, T. K., and V. S. Dayal. 1984. Ototoxicity and noise-drug interaction. J. Otolaryngol. 13:361–366.

Bidstrup, P. L. 1964. Toxicity of Mercury and its Compounds. Amsterdam: Elsevier Scientific Publishing Company.

Billmaier, D., H. T. Yee, N. Allen, B. Craft, N. Williams, S. Epstein, and R. Fontaine. 1974. Peripheral neuropathy in a coated fabrics plant. J. Occup. Med. 16:665–671.

Blair, V. W., A. R. Hollenbeck, R. F. Smith, and J. W. Scanlon. 1984. Neonatal preference for visual patterns: Modification by prenatal anesthetic exposure? Dev. Med. Child. Neurol. 26:476–483.

Boettcher, F. A., D. Henderson, M. A. Gratton, R. W. Danielson, and C. D. Byrne. 1987. Synergistic interactions of noise and other ototraumatic agents. Ear Hear. 8:192–212.

Bogo, V., T. A. Hill, and R. W. Young. 1981. Comparison of accelerod and rotorod sensitivity in detecting ethanol- and acrylamide-induced performance decrement in rats: Review of experimental considerations of rotating systems. Neurotoxicology 2:765–787.

Boylan, J. J., J. L. Egle, P. S. Guzelian. 1978. Cholestyramine: Use as a new therapeutic approach for chlordecone (Kepone) poisoning. Science 199:893–895.

Broadwell, R. D. 1989. Transcytosis of macromolecules through the blood-brain barrier: A cell biological perspective and critical appraisal. Acta Neuropathol. (Berl.) 79:117–128.

Brydon, J. E., V. H. Morgenroth, III, A. Smith, and R. Visser. 1990. OECD's Work on Investigation of High Production Volume Chemicals. EXHVA/90.163/2.5. Paris: OECD (Organisation for Economic Co-operation and Development). 5 February 1990. 17 pp.

Buchthal, F., and R. Behse. 1979. Electrophysiology and nerve biopsy in men exposed to lead. Br. J. Ind. Med. 36:135–147.

Buelke-Sam, J., C. A. Kimmel, J. Adams, C. J. Nelson, C. V. Vorhees, D. C. Wright, V. St. Omer, B. A. Korol, R. E. Butcher, M. A. Geyer, J. F. Holson, C. L. Kutscher, and M. J. Wayner. 1985. Collaborative behavioral teratology study: Results. Neurobehav. Toxicol. Teratol. 7:591–624.

Butcher, R. E., and C. V. Vorhees. 1979. A preliminary test battery for investigation of the behavioral teratology of selected psychotropic drugs. Neurobehav. Toxicol. 1(Suppl. 1):207–212.

Cabe, P. A., and H. A. Tilson. 1978. The hindlimb extensor response: A method for assessing motor dysfunction in rats. Pharmacol. Biochem. Behav. 9:133–136.

Calne, D. B., and J. W. Langston. 1983. Aetiology of Parkinson's disease. Lancet 2(8365-66):1457–1459.

Calne, D. B., J. W. Langston, W. R. W. Martin, A. J. Stoessl, T. J. Ruth, M. J. Adam, B. D. Pate, and M. Schulzer. 1985. Positron emission tomography after MPTP: Observations relating to the cause of Parkinson's disease. Nature 317:246–248.

Calne, D. B., A. Eisen, E. McGeer, and P. Spencer. 1986. Alzheimer's disease, Parkinson's disease, and motoneurone disease: A biotrophic interaction between ageing and environment? Lancet II(8515):1067–1070.

Cannon, S. B., J. M. Veazey, Jr., R. S. Jackson, V. W. Burse, C. Hayes, W. E. Straub, P. J. Landrigan, and J. A. Liddle. 1978. Epidemic kepone poisoning in chemical workers. Am. J. Epidemiol. 107:529—537.

Cassells, D. A. K., and E. C. Dodds. 1946. Tetra-ethyl lead poisoning. Brit. Med. J. 2:4479—4483.

Cavanagh, J. B. 1964. The significance of the "dying-back" process in experimental and human neurological disease. Int. Rev. Exp. Path. 3:219—267.

Cavanagh, J. B., and R. J. Bennetts. 1981. On the pattern of changes in the rat nervous system produced by 2,5-hexanediol: A topographical study by light microscopy. Brain 104:297—318.

Cavanagh, J. B., and F. C.-K. Chen. 1971. The effects of methyl-mercury-dicyandiamide on the peripheral nerves and spinal cord of rats. Acta Neuropathol. (Berlin) 19:208—215.

Cavanagh, J. B., and J. M. Jacobs. 1964. Some quantitative aspects of diphtheritic neuropathy. Brit. J. Exp. Path. 45:309—327.

CDC (Centers for Disease Control, U. S. Public Health Service). 1978. Acute and possible long-term effects of 1,3 dichloropropene—California. MMWR (Morbidity and Mortality Weekly Report) Feb. 17:50,55. U.S. Department of Health and Human Services. Washington, D.C.: U.S. Government Printing Office.

Chernoff, N., and Kavlock, R. J. 1982. An in vivo teratology screen utilizing pregnant mice. J. Toxicol. Environ. Health 10:541—550.

Chiba S., and K. Ando. 1976. Effects of chronic administration of kanamycin on conditioned suppression to auditory stimulus in rats. Jpn. J. Pharmacol. 26:419—426.

Cho, E-S., P. S. Spencer, and B. S. Jortner. 1980. Doxorubicin. Pp. 430—439 in Experimental and Clinical Neurotoxicology, P. S. Spencer and H. H. Schaumburg, eds. Baltimore: Williams & Wilkins.

Cohrssen, J. J., and V. T. Covello. 1989. Risk Analysis: A Guide to Principles and Methods for Analyzing Health and Environmental Risks. United States Council on Environmental Quality, Executive Office of the President. Available from NTIS as PB 89-137772.

Committee on Biological Markers of the National Research Council. 1987. Biological markers in environmental health research. Environ. Health Perspect. 74:3—9.

Cook, D. G., S. Fahn, and K. A. Brait. 1974. Chronic manganese intoxication. Arch. Neurol. 30:59—71.

Cossa, P., J. Duplay, L. Fishgold, L. Arfel-Capdeville, M. Passouant, and J. Radermecker. 1959. Encephalopathies toxiques au Stalinon: Aspects anatomocliniques et electroencephalographiques. Acta Neurol. Psychiatrica Belg. 59:281—303.

Couri, D., and J. P. Nachtman. 1979. Biochemical and biophysical studies of 2,5-hexanedione neuropathy. Neurotoxicology 1:269—283.

Cranmer, J. M., and L. Goldberg, eds. 1986. Proceedings of the workshop on neurobehavioral effects of solvents. Neurotoxicology 7:1—123.

Crofton, K. M., and L. P. Sheets. 1989. Evaluation of sensory system function using reflex modification of the startle response. J. Amer. Coll. Toxicol. 8:199—212.

Crouch, E. A., and R. Wilson. 1987. Risk assessment and comparisons: An introduction. Science: 236:267—270.

Crump, K. S. 1984. A new method for determining allowable daily intakes. Fundam. Appl. Toxicol. 4:854—871.

Cushner, I. M. 1981. Maternal behavior and perinatal risks: Alcohol, smoking, and drugs. Ann. Rev. Pub. Health 2:201—218.

Damstra, T., and S. C. Bondy. 1980. The current status and future of biochemical assays for neurotoxicity. Pp. 820—833 in Experimental and Clinical Neurotoxicology, P. S. Spencer and H. H. Schaumburg, eds. Baltimore: Williams & Wilkens.

Davies, P. W. 1968. The action potential. Pp. 1094—1120 in Medical Physiology, Vol. II, 12th ed., V. B. Mountcastle, ed. St. Louis: C. V. Mosby.

DeCaprio, A. P. 1985. Molecular mechanisms of diketone neurotoxicity. Chem. Biol. Interact. 54:257—270.

DeCaprio, A. P., E. J. Olajos, and P. Weber. 1982. Covalent binding of a neurotoxic *n*-hexane metabolite: Conversion of primary amines to substituted pyrrole adducts by 2,5-hexanedione. Toxicol. Appl. Pharmacol. 65:440—450.

DeCaprio, A. P., R. G. Briggs, S. J. Jackowski, and J. C. S. Kim. 1988. Comparative neurotoxicity and pyrrole-forming potential of 2,5-hexanedione and perdeuterio-2,5-hexanedione in the rat. Toxicol. Appl. Pharmacol. 92:75—85.

Denny-Brown, D. 1947. Neurological conditions resulting from prolonged and severe dietary restriction. Medicine 26:41—113.

Dick, R. B., and B. L. Johnson. 1986. Human experimental studies. Pp. 348—387 in Neurobehavioral Toxicology, Z. Annau, ed. Baltimore: The Johns Hopkins University Press.

Diener, R. M. 1987. Behavioral toxicology: Current industrial viewpoint. J. Am. Coll. Toxicol. 4:427—432.

Doull, J. 1980. Factors influencing toxicology. Pp. 70—83 in Casarett and Doull's Toxicology: Basic Science of Poisons, 2nd ed., J. Doull, C. D. Klaassen, and M. O. Amdur, eds. New York: Macmillan.

Dourson, M. L., and J. F. Stara. 1983. Regulatory history and experimental support of uncertainty (safety) factors. Regul. Toxicol. Pharmacol. 3:224—238.

Eckerman, D. A., J. B. Carroll, D. Foree, C. M. Gullion, M. Lansman, E. R. Long, M.

B. Waller, and T. S. Wallsten. 1985. An approach to brief field testing for neurotoxicity. Neurobehav. Toxicol. Teratol. 7:387—393.

Edelman, G. M. 1987. Neural Darwinism: The Theory of Neuronal Group Selection. New York: Basic Books. 371 pp.

Edwards, P. M., and V. H. Parker. 1977. A simple, sensitive and objective method for early assessment of acrylamide neuropathy in rats. Toxicol. Appl. Pharmacol. 40:589—591.

EPA (U.S. Environmental Protection Agency). 1985. Toxic Substances Control Act Test Guidelines. Final Rules. 40 CFR Part 798, Subpart 798.6050. FR 50(188):39458—39460.

Evans, H. L. 1989. Behaviors in the home cage reveal neurotoxicity: Recent findings and proposals for the future. J. Am. Coll. Toxicol. 8:35—52.

Evans, H. L., V. G. Laties, and B. Weiss. 1975. Behavioral effects of mercury and methylmercury. Fed. Proc. 34:1858—1867.

FDA (U.S. Food and Drug Administration). 1988. Good laboratory practice for nonclinical laboratory studies. Code of Federal Regulations, Title 21, Chap. 1, Pt. 58. Washington, D.C.: U.S. Government Printing Office.

Fechter, L. D., and J. S. Young. 1983. Discrimination of auditory from nonauditory toxicity by reflex modulation audiometry: Effects of triethyltin. Toxicol. Appl. Pharmacol. 70:216—227.

Federal Register. 1978. Pesticide Programs. Proposed Guidelines for Registering Pesticides in the U. S.; Hazard Evaluation: Humans and Domestic Animals. FR 43(163):37362. Washington, D.C.: U.S. Government Printing Office. August 22.

Federal Register. 1985. Toxic Substances Control Act Test Guidelines: Final Rule, Part II. Vol. FR 50(188):39458—39470. Washington, D.C.: U.S. Government Printing Office.

Fisher, F. 1980. Neurotoxicology and governmental regulation of chemicals in the United States. Pp. 874—882 in Experimental and Clinical Neurotoxicology, P. S. Spencer and H. H. Schaumburg, eds. Baltimore, Md.: Williams & Wilkins.

Foege, W. H., R. C. Hogan, and L. H. Newton. 1976. Surveillance projects for selected diseases. Int. J. Epidemiol. 5:29—37.

Fortemps, E., G. Amand, A. Bomboir, R. Lauwerys, and E. C. Laterre. 1978. Trimethyltin poisoning: A report of two cases. Int. Arch. Occup. Environ. Health 41:1—6.

Fowler, S. C. 1987. Force and duration of operant responses as dependent variables in behavioral pharmacology. Pp. 83—127 in Neurobehavioral Pharmacology, Vol. 6, Advances in Behavioral Pharmacology, T. Thompson, P. B. Dews, and J. E. Barrett, eds. New York: Erlbaum Press.

Fowler, B. A., and E. K. Silbergeld. 1989. Occupational diseases--new workforces, new workplaces. Ann. N. Y. Acad. Sci. 572:46—54.

Frazier, J. M. 1990. Validation of *in vitro* models. J. Am. Coll. Toxicol. 9:355—359.

French, L. R., L. M. Schuman, J. A. Mortimer, J. T. Hutton, R. A. Boatman, and B. Christians. 1985. Am. J. Epidemiol. 121:414—421.

Friberg, L. 1977. Toxicology of Metals, Vol. 2. Permanent Commission and International Association of Occupational Health in cooperation with the Swedish Environmental Protection Board and the Karolinska Institute. EPA-600/1-77-022. Report available through the NTIS as PB-268-324.

Friedland, R. P., W. J. Jagust, R. H. Huesman, E. Koss, B. Knittel, C. A. Mathis, B. A. Ober, B. M. Mazoyer, and T. F. Butinger. 1989. Regional cerebral glucose transport and utilization in Alzheimer's disease. Neurology 39:1427—1434.

Friedlander, T. R., and F. T. Hearne. 1980. Epidemiologic considerations in studying neurotoxic disorders. Pp. 650—662 in

Experimental and Clinical Neurotoxicology, P. S. Spencer and H. H. Schaumburg, eds. Baltimore: Williams & Wilkins.

Fromm, C., and E.V. Evarts. 1981. Relation of size and activity of motor cortex pyramidal tract neurons during skilled movements in the monkey. J. Neurosci. 1:453—460.

Fullerton, P. M. 1966. Chronic peripheral neuropathy produced by lead poisoning in guinea-pigs. J. Neuropathol. Exp. Neurol. 25:214—236.

Fullerton, P. M. 1969. Electrophysiological and histological observations on peripheral nerves in acrylamide poisoning in man. J. Neurol. Neurosurg. Psychiatry 32:186—192.

Gable, C. B. 1990. A compendium of public health data sources. Am. J. Epidemiology 131:381—393.

Gad, S. C. 1982. A neuromuscular screen for use in industrial toxicology. J. Toxicol. Environ. Health 9:691—704.

Gad, S. C., J. A. McKelvey, and R. A. Turney. 1979. NIAX Catalyst ESN: Subchronic neuropharmacology and neurotoxicology. Drug Chem. Toxicol. 2:223—236.

Gade, A., E. L. Mortensen, and P. Bruhn. 1988. "Chronic painter's syndrome": A reanalysis of psychological test data in a group of diagnosed cases, based on comparisons with matched controls. Acta Neurol. Scand. 77:293—306.

GAO (General Accounting Office). 1990a. Toxic Substances: EPA's Chemical Testing Program Has Made Little Progress. Report to the Chairman, Subcommittee on Environment, Energy, and Natural Resources, Committee on Government Operations, U.S. House of Representatives. GAO/RCED-90-112. U.S. General Accounting Office, Washington, D.C. April, 1990.

GAO (General Accounting Office). 1990b. Toxic Substances: Effectiveness of Unreasonable Risk Standards Unclear.

Report to the Chairman, Subcommittee on Health and the Environment, Committee on Energy and Commerce, U.S. House of Representatives. GAO/RCED-90-189. U.S. General Accounting Office, Washington, D.C. July, 1990.

Gaylor, D. W. 1983. The use of safety factors for controlling risk. J. Toxicol. Environ. Health 11:329—336.

Gaylor, D. W., and W. Slikker, Jr. 1990. Risk assessment for neurotoxic effects. Neurotoxicology 11:211—218.

Genter, M. B., G. Szakál-Quin, C. W. Anderson, D. C. Anthony, and D. G. Graham. 1987. Evidence that pyrrole formation is a pathogenetic step in γ-diketone neuropathy. Toxicol. Appl. Pharmacol. 87:351—362.

Gerhart. K. M., J. S. Hong, and H. A. Tilson. 1985. Studies on the mechanism of chlordecone-induced tremor in rats. Neurotoxicology 2:765—787.

Gibaldi, M., and D. Perrier. 1982. Pharmacokinetics, 2nd ed. New York: Marcel Dekker. 494 pp.

Goetz, C. G. 1985. Pesticides and other environmental toxins. Pp. 107—131 in Neurotoxins in Clinical Practice. New York: Spectrum Publications, Inc.

Goldman, P. S. 1971. Functional development of the prefrontal cortex in early life and the problem of neuronal plasticity. Exp. Neurol. 32:366—387.

Goldman, R. H., and J. M. Peters. 1981. The occupational and environmental health history. JAMA 246:2831—2836.

Goldstein, A., and H. Kalant. 1990. Drug policy: Striking the right balance. Science 249:1513—1521.

Graham, D. G. 1980. Hexane neuropathy: A proposal for pathogenesis of a hazard of occupational exposure and inhalant abuse. Chem. Biol. Interact. 32:339—345.

Graham, D. I., S. U. Kim, N. K. Gonatas, and L. Guyotte. 1975. The neurotoxic effects of triethyltin (TET) sulfate on myelinating cultures of mouse spinal cord. J. Neuropath. Exp. Neurol. 34:401—412.

Graham, D. G., S. M. Tiffany, W. R. Bell, Jr., and W. F. Gutknecht. 1978. Autooxidation versus covalent binding of quinones as the mechanism of toxicity of dopamine, 6-hydroxydopamine and related compounds for C1300 neuroblastoma cells in vitro. Mol. Pharmacol. 14:644—653.

Graham, D. G., D. C. Anthony, K. Boelkelheide, N. A. Maschmann, R. G. Richards, J. W. Wolfram, and B. R. Shaw. 1982. Studies of the molecular pathogenesis of hexane neuropathy. II. Evidence that pyrrole derivatization of lysyl residues leads to protein crosslinking. Toxicol. Appl. Pharmacol. 64:415—422.

Griffin, J. W., and D. L. Price. 1980. Proximal axonopathies induced by toxic chemicals. Pp. 161—178 in Experimental and Clinical Neurotoxicology, P. S. Spencer and H. H. Schaumburg, eds. Baltimore, Md.: Williams & Wilkins.

Guzelian, P. S. 1982. Comparative toxicology of chlordecone (Kepone) in humans and experimental animals. Annu. Rev. Pharmacol. Toxicol. 22:89—113.

Haggerty, G. C. 1989. Development of Tier 1 neurobehavioral capabilities for incorporation into pivotal rodent safety assessment studies. J. Amer. Coll. Toxicol. 8:53—69.

Hammerschlag, R., and S. T. Brady. 1989. Axonal transport and the neuronal cytoskeleton. Pp. 457—478 in Basic Neurochemistry, 4th ed., G. Siegel, B. Agranoff, R. W. Albers, and P. Molinoff, eds. New York: Raven Press.

Hänninen, H., and K. Lindström. 1979. Behavioral Test Battery for Toxicopsychological Studies: Used at the Institute of Occupational Health in Helsinki, 2nd ed. Helsinki, Finland: Institute of Occupational Health.

Hattis, D., and K. Shapiro. 1990. Analysis of dose/time/response relationships for chronic toxic effects: The case of acrylamide. Neurotoxicology 11:219—236.

Heise, G. A. 1984. Behavioral methods for

measuring effects of drugs on learning and memory in animals. Med. Res. Rev. 4:535—558.

Herskowitz, A., N. Ishii, and H. Schaumburg. 1971. n-hexane neuropathy: A syndrome occurring as a result of industrial exposure. N. Engl. J. Med. 285:82—85.

Hertzman, C., M. Wiens, D. Bowering, B. Snow, and D. Calne. 1990. Parkinson's disease: A case-control study of occupational and environmental risk factors. Am J. Indust. Med. 17:349—355.

Hierons, R., and M. K. Johnson. 1978. Clinical and toxicological investigations of a case of delayed neuropathy in man after acute poisoning by an organophosphorus pesticide. Arch. Toxicol. 40:279—284.

Hill, R. M., and L. M. Tennyson. 1986. Maternal drug therapy: Effect on fetal and neonatal growth and neurobehavior. Neurotoxicology 7:121—140.

Hille, B. 1984. Ionic Channels of Excitable Membranes. Sunderland, Mass.: Sinauer Associates.

Hopkins, A. P., and Gilliatt, R. W. 1971. Motor sensory nerve conduction velocity in the baboon: Normal values and changes during acrylamide neuropathy. J. Neurol. Neurosurg. Psychiatry 34:415—426.

Horan, J. M., T. L. Kurt, P. J. Landrigan, J. M. Melius, and M. Singal. 1985. Neurologic dysfunction from exposure to 2-t-butylazo-2-hydroxy-5-methylhexane (BHMH): A new occupational neuropathy. Am. J. Pub. Health 75:513—517.

Hornykiewicz, O. 1986. Dopamine deficiency and dopamine substitution in Parkinson's disease. Pp. 319—330 in The Neurobiology of Dopamine Systems, W. Winlow and R. Markstein, eds. Manchester: Manchester University Press.

Hruska, R.E., S. Kennedy, and E.K. Silbergeld. 1979. Quantitative aspects of normal locomotion in rats. Life Sci. 25:171—179.

Hubel, D. H., and M. S. Livingstone. 1987. Segregation of form, color, and stereopsis in primate area 18.   J. Neurosci. 7:3378—3415.

Hunter, D., and D. S. Russell. 1954. Focal cerebral and cerebellar atrophy in human subjects due to organic mercury compounds. J. Neurol. Neurosurg. Psychiatr. 17:235—241.

IOM (Institute of Medicine). 1991. Seafood Safety. Washington, D.C.: National Academy Press. 446 pp.

Irwin, S. 1968. Comprehensive observational assessment: Ia. A systematic quantitative procedure for assessing the behavioral and physiologic state of a mouse. Psychopharmacologia 13:222—257.

Isaacson, R. L., A. J. Nonneman, and L. W. Schmalz. 1968. Behavioral and anatomical sequelae of damage to the infant limbic system. Pp. 41—78 in The Neuropsychology of Development. New York: Wiley.

Jenner, P. 1989. Clues to the mechanism underlying dopamine cell death in Parkinson's disease. J. Neurol. Neurosurg. Psychiatry. Special Supplement:22—28.

Johnson, B. L., ed. 1987. Prevention of Neurotoxic Illness in Working Populations. New York: John Wiley & Sons. 257 pp.

Johnson, B. L., and W. K. Anger. 1983. Behavioral toxicology. Pp. 329—350 in Environmental and Occupational Medicine, W. N. Rom, ed. Boston: Little, Brown and Company.

Johnson, D., K. Houghton, C. Siegel, J. Martyny, L. Cook, and E. J. Mangione. 1989. Occupational and paraoccupational exposure to lead. MMWR 38:338—340, 345. U.S. Department of Health and Human Services. Washington, D.C.: U.S. Government Printing Office.

Johnson, B. L., W. K. Anger, A. DuRao, and C. Xintaras, eds. 1990. Advances in Neurobehavioral Toxicology: Applications in Environmental and Occupational Health. Chelsea, Mich: Lewis Publishers, Inc.

Johnson, M. K. 1977. Improved assay of

neurotoxic esterase for screening organophosphates for delayed neurotoxicity potential. Arch. Toxicol. 37:113—115.

Jones, H. B., and J. B. Cavanagh. 1983. Distortions of the nodes of Ranvier from axonal distension by filamentous masses in hexacarbon intoxication. J. Neurocytol. 12:439—458.

Jones, K. L., and D. W. Smith. 1973. Recognition of the fetal alcohol syndrome in early infancy. Lancet 2:999—1001.

Katz, G. V., J. L. O'Donoghue, G. D. DiVincenzo, and C. J. Terhaar. 1980. Comparative neurotoxicity and metabolism of ethyl n-butyl ketone in rats. Toxicol. Appl. Pharmacol. 52:153—158.

Kellogg, C. K. 1985. Prenatal diazepam exposure in rats: Long-lasting functional changes in the offspring. Neurobehav. Toxicol. Teratol. 7:483—488.

Kelly, J. P. 1985. Reactions of neurons to injury. Pp. 187—195 in Principles of Neural Science, 2nd ed., E. R. Kandel and J. H. Schwartz, eds. New York: Elsevier Science Publishing Co., Inc.

Kim, S. U. 1971. Neurotoxic effects of alkyl mercury compound on myelinating cultures of mouse cerebellum. Exp. Neurol. 32:237—246.

Kimmel, C. A. 1990. Quantitative approaches to human risk assessment for noncancer health effects. Neurotoxicology 11:189—198.

Klaassen, C. D. 1986. Distribution, excretion, and absorption of toxicants. Pp. 33—63 in Casarett and Doull's Toxicology: The Basic Science of Poisons, 3rd ed., C. D. Klaassen, M. O. Amdur, and J. Doull, eds. New York: Plenum Press.

Kopin, I. J., and S. P. Markey. 1988. MPTP toxicity: Implications for research in Parkinson's disease. Annu. Rev. Neurosci. 11:81—96.

Korobkin, R., A. K. Asbury, A. J. Sumner, and S. L. Nielsen. 1975. Glue-sniffing neuropathy. Arch. Neurol. 32:158—162.

Krasavage, W. J., J. L. O'Donoghue, G. D. DiVincenzo, and C. J. Terhaar. 1980. The relative neurotoxicity of methyl-n-butyl ketone, n-hexane and their metabolites. Toxicol. Appl. Pharmacol. 52:433—441.

Krewski, D., K. S. Crump, J. Farmer, D. W. Gaylor, R. Howe, C. Portier, D. Salsburg, R. L. Sielken, and J. van Ryzin. 1983. A comparison of statistical methods for low dose extrapolation utilizing time-to-tumor data. Fundam. Appl. Toxicol. 3:140—160.

Kulig, B.M. 1989. A neurofunctional test battery for evaluating the effects of long-term exposure to chemicals. J. Amer. Coll. Toxicol. 8:71—83.

Kurland, L. T. 1963. Epidemiological investigations of neurological disorders in the Mariana Islands. Pp. 219—223 in Epidemiology Reports on Research and Teaching, J. Pemberton, ed. New York: Oxford University Press.

Landrigan, P. J. 1989. Improving the surveillance of occupational disease. Am. J. Public Health 79:1601—1602.

Langston, J. W., and I. Irwin. 1986. MPTP: Current concepts and controversies. Clin. Neuropharmacol. 9:485—507.

Langston, J. W., P. Ballard, J. W. Tetrud, and I. Irwin. 1983. Chronic parkinsonism in humans due to a product of meperidine-analog synthesis. Science 219:979—980.

Lave, L.B., and G.S. Omenn. 1986. Cost-effectiveness of short-term tests for carcinogenicity. Nature 324:29—34.

Lawrence, L. J., and J. E. Casida. 1983. Stereospecific action of pyrethroid insecticides on the γ-aminobutyric acid receptor-ionophore complex. Science 221:1399—1401.

LeQuesne, P. M. 1980. Acrylamide. Pp. 309—325 in Clinical and Experimental Neurotoxicology, P. S. Spencer, and H. H. Schaumburg, eds. Baltimore: Williams & Wilkins.

LeQuesne, P. M. 1987. Clinically used electrophysiological end-points. Pp. 103—116 in Electrophysiology in Neurotoxicology, Vol. 1, H. E. Lowndes, ed.

Boca Raton, Fla.: CRC Press.

LeQuesne, P. M., and J. G. McLeod. 1977. Peripheral neuropathy following a single exposure to arsenic. Clinical course in four patients with electrophysiological and histological studies. J. Neurol. Sci. 32:437—451.

Letz, R. 1991. Use of computerized test batteries for quantifying neurobehavioral outcomes. Environ. Health Perspect. 90:195—198.

Letz, R., and E. L. Baker. 1986. Computer-administered neurobehavioral testing in occupational health. Sem. Occup. Med. 1:197-203.

Liebman, M. 1991. Neuroanatomy Made Easy and Understandable, 4th ed. Gaithersburg, Md.: Aspen Publishers, Inc.

Lilienfeld, D. E., E. Chan, J. Ehland, J. Godbold, P. J. Landrigan, G. Marsh, and D. P. Perl. 1989. Rising mortality from motoneuron disease in the USA, 1962-84. Lancet 1(8640):710—713.

Lim, D. J. 1986. Effects of noise and ototoxic drugs at the cellular level in the cochlea: A review. Am. J. Otolaryngol. 7:73—99.

Little, R. E., K. W. Anderson, C. H. Ervin, B. Worthington-Roberts, and S. K. Clarren. 1989. Maternal alcohol use during breast-feeding and infant mental and motor development at one year. N. Engl. J. Med. 321:425—430.

Lowndes, H. E., and T. Baker. 1976. Studies on drug-induced neuropathies. III. Motor nerve deficit in cats with experimental acrylamide neuropathy. Eur. J. Pharmacol. 35:177—184.

Lowndes, H. E., T. Baker, E. S. Cho, and B. S. Jortner. 1978a. Position sensitivity of de-efferented muscle spindles in experimental acrylamide neuropathy. J. Pharmacol. Exp. Ther. 205:40—48.

Lowndes, H. E., T. Baker, L. P. Michelson, and M. Vincent-Ablazey. 1978b. Attenuated dynamic responses of primary endings of muscle spindles: A basis for depressed tendon responses in acrylamide

neuropathy. Ann. Nurol. 3:433—437.

Lund, A. E., and T. Narahashi. 1982. Dose-dependent interaction of the pyrethroid isomers with sodium channels of squid axon membranes. Neurotoxicol. 3:11—24.

Lund, A. E., and T. Narahashi. 1983. Kinetics of sodium channel modification as the basis for the variation in the nerve membrane effects of pyrethroids and DDT analogs. Pestic. Biochem. Physiol. 20:203—216.

MacMahon, B., and T. F. Pugh. 1970. Epidemiology: Principles and Methods. Boston: Little, Brown and Co. 376 pp.

MacPhail, R. C. 1985. Effects of pesticides on schedule-controlled behavior. Pp. 519—535 in Behavioral Pharmacology: The Current Status, L. S. Seiden and R. L. Balster, eds. New York: Alan R. Liss. 571 pp.

MacPhail, R. C., D. B. Peele, and K. M. Crofton. 1989. Motor activity and screening for neurotoxicity. J. Amer. Coll. Toxicol. 8:117—125.

Mailman, R. B. 1987. Mechanisms of CNS injury in behavioral dysfunction. Neurotoxicol. Teratol. 9:417—426.

Marsh, D. O., T. W. Clarkson, C. Cox, G. J. Myers, L. Amin-Zaki, and S. Al-Tikriti. 1987. Fetal methylmercury poisoning. Relationship between concentration in single strands of maternal hair and child effects. Arch. Neurol 44:1017—1022.

Martin-Bouyer, G., R. Lebreton, M. Toga, P. D. Stolley, and J. Lockhart. 1982. Outbreak of accidental hexachlorophene poisoning in France. Lancet 1(8263):91—95.

Marteniuk, G. 1976. Information Processing in Motor Skills. New York: Holt, Rinehart, and Winston. 244 pp.

Marwaha, J., and W. J. Anderson, eds. 1984. Neuroreceptors in Health and Disease. New York: Karger. 255 pp.

Matsumoto, H. G., D. Koya, and T. Takeuchi. 1965. Fetal Minamata disease: A neuropathological study of two cases of intrauterine intoxication by a methylmercury compound. J. Neuropathol. Exp. Neurol. 24:563—574.

Mattsson, J. L., and R. R. Albee. 1988. Sensory evoked potentials in neurotoxicology. Neurotoxicol. Teratol. 10:435—443.

Mattson, J. L., R. R. Albee, and D. L. Eisenbrandt. 1989. Neurological approach to neurotoxicological evaluation in laboratory animals. J. Amer. Coll. Toxicol. 8:271—286.

Maurissen, J. P. J. 1988. Quantitative sensory assessment in toxicology and occupational medicine: Applications, theory, and critical appraisal. Toxicol. Lett. 43:321—343.

Maurissen, J. P. J., and J. L. Mattsson. 1989. Critical assessment of motor activity as a screen for neurotoxicity. Toxicol. Indust. Health 5:195—202.

Maurissen, J. P. J., B. Weiss, and H. T. Davis. 1983. Somatosensory thresholds in monkeys exposed to acrylamide. Toxicol. Appl. Pharmacol. 71:266-279.

McGeer, P. L., E. G. McGeer, and J. S. Suzuki. 1977. Aging and extrapyramidal function. Arch. Neurol. 34:33—35.

McKeown-Eyssen, G. E., J. Ruedy, and A. Neims. 1983. Methylmercury exposure in northern Quebec. II. Neurologic findings in children. Am. J. Epidem. 118:470—479.

Merigan, W. H., E. Barkdoll, and J. P. J. Maurissen. 1982. Acrylamide-induced visual impairment in primates. Toxicol. Appl. Pharmacol. 62:342—345.

Meyer, O. A., H. A. Tilson, W. C. Bryd, and M. T. Riley. 1979. A method for the routine assessment of fore- and hind-limb grip strength of rats and mice. Neurobehav. Toxicol. 1:233—236.

Miller, J. C., and A. J. Friedhoff. 1986. Prenatal neuroleptic exposure alters postnatal striatal cholinergic activity in the rat. Dev. Neurosci. 8:111—116.

Miller, M. S., and P. S. Spencer. 1985. The mechanisms of acrylamide axonopathy. Annu. Rev. Pharmacol. Toxicol. 25:643—666.

Milnor, W. R. 1968. Blood supply of special regions. Pp. 221—243 in Medical Physiology, Vol. I, 12th ed., V. B. Mountcastle, ed. St. Louis: C. V. Mosby.

Misumi, J., and M. Nagano. 1984. Neurophysiological studies on the relation between the structural properties and neurotoxicity of aliphatic hydrocarbon compounds in rats. Br. J. Ind. Med. 41:526—532.

MMWR (Morbidity and Mortality Weekly Report). 1986. Aldicarb food poisoning from contaminated melons—California. JAMA 256:175—176.

Moore, R. W., C. R. Jefcoate, and R. E. Peterson. 1991. 2,3,7,8-Tetrachlorodibenzo-p-dioxin inhibits steroidogenesis in the rat testis by inhibiting the mobilization of cholesterol to cytochrome P450scc. Toxicol. Appl. Pharmacol. 109:85—97.

Moser, V. C. 1989. Screening approaches to neurotoxicity: A functional observational battery. J. Amer. Coll. Toxicol. 8:85—94.

Moser, V. C., J. P. McCormick, J. P. Creason, and R. C. MacPhail. 1988. Comparison of chlordimeform and carbaryl using a functional observational battery. Fundam. Appl. Toxicol. 11:189—206.

Murai, Y., S. Shiraishi, Y. Yamashita, A. Ohnishi, and K. Arimura. 1982. Neurophysiological effects of methylmercury on the nervous system. Electroencephalog. Clin. Neurophysiol. Suppl. 36:682—687.

Narahashi, T. 1971. Effects of insecticides on excitable tissues. Pp. 1—93 in Advances in Insect Physiology, Vol. 8, J. W. L. Beament, J. E. Treherne, and V. B. Wigglesworth, eds. London and New York: Academic Press.

Narahashi, T. 1985. Nerve membrane ionic channels as the primary target of pyrethroids. Neurotoxicology 6:3—22.

Narahashi, T. 1989. The role of ion channels in insecticides. Pp. 55—84 in Insecticide Action: From Molecule to Organism, T. Narahashi and J. E. Chambers, eds. New York: Plenum Press.

Narahashi, T., and J. M. Frey. 1989. Lindane and cyclodiene insecticides block GABA-activated chloride current in cultured rat hippocampal neurons. Paper presented at the 19th annual meeting of the Society for Neuroscience, Phoenix, Ariz. 29 October–3 November.

Narahashi, T., and A. E. Lund. 1980. Giant axons as models for the study of the mechanism of action of insecticides. Pp. 497–505 in Insect Neurobiology and Pesticide Action (Neurotox 79). Proceedings of a Society of Chemical Industry Symposium, York, England, 3-7 September, 1979. London: Society of Chemical Industry.

National Conference on Clustering of Health Events. 1990. Am. J. Epidem. 132(1):Supple.

NCCLS (National Committee for Clinical Laboratory Standards). 1981. List of Standards. Villanova, Penn.: National Committee for Clinical Laboratory Standards.

NCCLS (National Committee for Clinical Laboratory Standards). 1985. List of Standards. Villanova, Penn.: National Committee for Clinical Laboratory Standards.

NCHS (National Center for Health Statistics). 1984. Blood lead levels for persons age 6 months-74 years: United States, 1976-80. Data from the National Health and Nutrition Examination Survey. Vital and Health Statistics, Series 11, No. 233. DHHS Pub. No. (PHS) 84-1683. Public Health Service, Washington, D.C.: U.S. Government Printing Office.

Needleman, H. L. 1986. Epidemiological studies. Pp. 279–287 in Neurobehavioral Toxicology, Z. Annau, ed. Baltimore: The Johns Hopkins University Press.

Needleman, H. L. 1989. The persistent threat of lead: A singular opportunity. Commentaries in Am. J. Pub. Health 79:643–645.

Needleman, H. L., and C. A. Gatsonis. 1990. Low-level lead exposure and the IQ of children: A meta-analysis of modern studies. JAMA 263:673–678.

Newland, M. C. 1988. Quantification of motor function in toxicology. Toxicol. Lett. 43:295–319. Elsevier Science Publishers.

NIEHS (National Institute for Environmental Health Sciences), Task Force 3. 1985. Biochemical and Cellular Markers of Chemical Exposure and Preclinical Indicators of Disease. Washington, D.C.: U.S. Department of Health and Human Services.

NIOSH (National Institute for Occupational Safety and Health). 1977. National Occupational Hazard Survey, Vol. 3, Survey Analysis and Supplemental Tables. USDHEW (NIOSH) Pub. No. 78-114. Cincinnati, Ohio: NIOSH Publications Office.

NIOSH (National Institute for Occupational Safety and Health). 1986. Leading work-related diseases and injuries--United States. MMWR 35:113–116, 121–123.

NRC (National Research Council). 1977. Principles and Procedures for Evaluating the Toxicity of Household Substances. Washington, D.C.: National Academy Press. 140 pp.

NRC (National Research Council). 1983. Risk Assessment in the Federal Government: Managing the Process. Washington D.C.: National Academy Press. 191 pp.

NRC (National Research Council). 1984. Toxicity Testing: Strategies to Determine Needs and Priorities. Washington D.C.: National Academy Press. 382 pp.

NRC (National Research Council). 1986. Drinking Water and Health. Vol. 6. Washington D.C.: National Academy Press. 457 pp.

NRC (National Research Council). 1987a. Counting Injuries and Illnesses in the Workplace. Washington, D.C.: National Academy Press. 176 pp.

NRC (National Research Council). 1987b. Drinking Water and Health. Vol. 8,

Pharmacokinetics in Risk Assessment. Washington, D.C.: National Academy Press. 488 pp.

NRC (National Research Council). 1989a. Biologic Markers in Pulmonary Toxicology. Washington, D.C.: National Academy Press. 179 pp.

NRC (National Research Council). 1989b. Biologic Markers in Reproductive Toxicology. Washington, D.C.: National Academy Press. 395 pp.

NRC (National Research Council). 1990. Health Effects of Exposure to Low Levels of Ionizing Radiation (BIER 5). Washington, D.C.: National Academy Press. 433 pp.

NRC (National Research Council). 1991a. Biological Markers of Immunotoxicology. Washington, D.C.: National Academy Press. (In press.)

NRC (National Research Council). 1991b. Monitoring Human Tissues for Toxic Substances. Washington, D.C.: National Academy Press. 211 pp.

Nriagu, J. O. 1978. Properties and the biogeochemical cycle of lead. Pp. 1—4 in the Biogeochemistry of Lead in the Environment, Part A, J. J. Nriagu, ed. Amsterdam: Elsevier/North Holland Biomedical Press.

O'Callaghan, J. P. 1988. Neurotypic and gliotypic proteins as biochemical markers of neurotoxicity. Neurotoxicol. Teratol. 10:445—452. New York, N.Y.: Pergamon Press.

O'Donoghue, J. L. 1986. Neurotoxic Risk Assessment in Industry. Paper presented at the annual meeting of the American Association for the Advancement of Science symposium Evaluating Neurotoxic Risk Posed in the Workplace and Environment, Philadelphia, Pa., May 26.

O'Donoghue, J. L. 1989. Screening for neurotoxicity using a neurologically based examination and neuropathology. J. Amer. Coll. Toxicol 8:97—115.

Ogata, N., S. M. Vogel, and T. Narahashi. 1988. Lindane but not deltamethrin

blocks a component of GABA-activated chloride channels. Fed. Am. Soc. Exp. Biol. J. (FASEB) 2:2895—2900.

Omenn, G. S. 1986. Susceptibility to occupational and environmental exposure to chemicals. Prog. Clin. Biol. Res. 214:527—545.

OTA (Office of Technology Assessment), U.S. Congress. 1984. Impacts of Neuroscience: A Background Paper. OTA-BP--BA-24. Washington, D.C.: U.S. Government Printing Office.

OTA (Office of Technology Assessment), U.S. Congress. 1987. Identifying and Regulating Carcinogens. OTA-BP-H-42. Washington, D.C.: U.S. Government Printing Office.

OTA (Office of Technology Assessment), U.S. Congress. 1990. Neurotoxicity: Identifying and Controlling Poisons of the Nervous System, OTA-BA-436. Washington, D.C.: U.S. Government Printing Office.

Otake, M., and W. J. Schull. 1984. In utero exposure to A-bomb radiation and mental retardation: A reassessment. Br. J. Radiol. 57:409—414.

Otto, D., and D. Eckerman, eds. 1985. Workshop on neurotoxicity testing in human populations: Workshop overview. Neurobehav. Toxicol. Teratol. 7:283-418.

Overstreet, D. H. 1977. Pharmacological approaches to habituation of the acoustic startle response in rats. Physiol. Psychol. 5:230—238.

Pardridge, W. M. 1988. Recent advances in blood-brain barrier transport. Annu. Rev. Pharmacol. Toxicol. 28:25—39.

Parkinson, D. K., E. J. Bromet, S. Cohen, L. O. Dunn, M. A. Dew, C. Ryan, and J. E. Schwartz. 1990. Health effects of long-term solvent exposure among women in blue-collar occupations. Am. J. Ind. Med.17:661—675.

Pearson, D. T., and K. N. Dietrich. 1985. The behavioral toxicology and teratology of childhood: Models, methods, and implications for intervention. Neurotoxicology 6:165—182.

Peele, D. B. 1989. Learning and memory: Considerations for toxicology. J. Amer. Coll. Toxicol. 8:213—224.

Perl, T. M., L. Bedard, T. Kosatsky, J. C. Hockin, E. D. C. Todd, and R. S. Remis. 1990. An outbreak of toxic encephalopathy caused by eating mussels contaminated with domoic acid. N. Engl. J. Med. 322:1775—1780.

Pestronk, A., J. P. Keogh, and J. W. Griffin. 1979. Dimethylaminopropionitrile (DMAPN) intoxication: A new industrial neuropathy. Neurology 29:540.

Peto, R., M. C. Pike, N. E. Day, R. G. Gray, P. N. Lee, S. Parish, J. Peto, S. Richards, and J. Wahrendorf. 1980. Guidelines for simple, sensitive significance tests for carcinogenic effects in long-term animal experiments. Monographs on the Longterm and Short-term Screening Assays for Carcinogens: A Critical Appraisal. International Agency for Research Against Cancer (IARC). Supplement 2:311—426. Geneva: World Health Organization.

Pierce, P. E., J. F. Thompson, W. H. Likosky, L. N. Nickey, W. F. Barthel, and A. R. Hinman. 1972. Alkyl mercury poisoning in humans. Report of an outbreak. JAMA 220:1439—1442.

Politis, M. J., H. H. Schaumburg, and P. S. Spencer. 1980. Neurotoxicity of selected chemicals. P. 619 in Experimental and Clinical Neurotoxicology, P. S. Spencer and H. H. Schaumburg, eds. Baltimore: Williams & Wilkins.

Provenzano, G. 1980. The social costs of excessive lead exposure during childhood. Pp. 199—315 in Low Level Lead Exposure, H. L. Needleman, ed. New York: Raven Press.

Pryor, G.T., J. Dickinson, R. A. Howd, and C. S. Rebert. 1983. Transient cognitive deficits and high-frequency hearing loss in weanling rats exposed to toluene. Neurobehav. Toxicol. Teratol. 5:53—57.

Rakic, P., and K. P. Riley. 1983. Overproduction and elimination of retinal axons in the fetal rhesus monkey. Science 219:1441—1444.

Rebert, C. S., S. S. Sorenson, R. A. Howd, and G. T. Pryor. 1983. Toluene-induced hearing loss in rats evidenced by the brainstem auditory-evoked response. Neurobehav. Toxicol. Teratol. 5:59—62.

Reiter, L. W. 1980. Neurotoxicology--meet the real world [editorial]. Neurobehav. Toxicol. 2:73—74.

Reiter, L. W. 1987. Neurotoxicology in regulation and risk assessment. Dev. Pharmacol. Ther. 10:354—368.

Reiter, L. W., and R. C. MacPhail. 1979. Motor activity: A survey of methods with potential use in toxicity testing. Neurobehav. Toxicol. 1(Suppl. 1):53—66.

Rice, D. C. 1988. Quantification of operant behavior. Toxicol. Lett. 43:361—379.

Rice, D. P., T. A. Hodgson, and A. N. Kopstein. 1985. The economic costs of illness: A replication and update. Health Care Financing Review 7:68—69. U.S. Department of Health and Human Services, Health Care Financing Administration, Baltimore, Md.

Riley, A. L., and D. L. Tuck. 1985. Conditioned taste aversions: A behavioral index of toxicity. Ann. N.Y. Acad. Sci. 443:272—292.

Rodier, P. M. 1986. Time of exposure and time of testing in developmental neurotoxicology. Neurotoxicology 7:69—76.

Rodier, P. M., W. S. Webster, and J. Langman. 1975. Morphological and behavioral consequences of chemically-induced lesions of the CNS. Pp. 169—176 in Aberrant Development of Human Infancy: Human and Animal Studies, N. Ellis, ed. New York: Erlbaum Press.

Rodier, P. M., B. Kates, W. A. White, and A. Muhs. 1991. Effects of prenatal exposure to methylazoxymethanol (MAM) on brain weight, hypothalamic cell number, pituitary structure, and postnatal growth in the rat. Teratol ogy 43:241—251.

Ronnett, G. V., L. D. Hester, J. S. Nye, K.

Connors, and S. H. Snyder. 1990. Human cortical neuronal cell line: Establishment from a patient with unilateral megalencephaly. Science 248:603—605.

Rosenberg, C. K., M. B. Genter, G. Szakál-Quin, D. C. Anthony, and D. G. Graham. 1987. *dl*-versus *meso*-3,4-dimethyl-2,5-hexanedione: A morphometric study of the proximo-distal distribution of axonal swellings in the anterior root of the rat. Toxicol. Appl. Pharmacol. 87:363—373.

Rosengarten, H., and A. J. Friedhoff. 1979. Enduring changes in dopamine receptor cells of pups from drug administration to pregnant and nursing rats. Science 203:1-133—1135.

Rosner, D., and G. Markowitz. 1985. A "gift of God"?: The public health controversy over leaded gasoline during the 1920s. Am. J. Public Health 75:344—52.

Ruigt, G. S. F. 1984. Pyrethroids. Ch. 7 in Comprehensive Insect Physiology, Biochemistry and Pharmacology, Vol. 12, G. A. Kerkut and L. I. Gilbert, eds. Oxford: Pergamon Press.

Rutstein, D. D., W. Berenberg, T. C. Chalmers, A. P. Fishman, and E. B. Perrin. 1976. Measuring the quality of medical care: A clinical method. N. Engl. J. Med. 294:582—588.

Rutstein, D. D., R. J. Mullan, T. M. Frazier, W. E. Halperin, J. M. Melius, and J. P. Sestito. 1983. Sentinel health events (occupational): A basis for physician recognition and public health surveillance. Am. J. Public Health 73:1054—1062.

Sager, P. R., M. Aschner, and P. M. Rodier. 1984. Persistent, differential alterations in developing cerebellar cortex of male and female mice after methylmercury exposure. Dev. Brain Res. 12:1—11.

Sayre, L. M., L. Autilio-Gambetti, and P. Gambetti. 1985. Pathogenesis of experimental giant neurofilamentous axonopathies: A unified hypothesis based on chemical modification of neurofilaments. Brain Res. Rev. 10:69—83.

Sayre, L. M., C. M. Shearson, T. Wong-

mongkolrit, R. Medori, and P. Gambetti. 1986. Structural basis of γ-diketone neurotoxicity: Non-neurotoxicity of 3,3-dimethyl-2,5-hexanedione, a γ-diketone incapable of pyrrole formation. Toxicol. Appl. Pharmacol. 84:36—44.

Schaumburg, H. H., and P. S. Spencer. 1976. Degeneration in central and peripheral nervous systems produced by pure *n*-hexane: An experimental study. Brain 99:183—192.

Schrot, J., J. R. Thomas, and R. A. Banvard. 1980. Modification of the repeated acquisition of response sequences in rats by low-level microwave exposure. Bioelectromagnetics 1:89—99.

Schrot, J., J. R. Thomas, and R. F. Robertson. 1984. Temporal changes in repeated acquisition behavior after carbon monoxide exposure. Neurobehav. Toxicol. Teratol. 6:23—28.

Schull, W. J., S. Norton, and R. P. Jensh. 1990. Ionizing radiation and the developing brain. Neurotoxicol. Teratol. 12:249—260.

Schwartz, J. H. 1985. Molecular aspects of post-synaptic receptors. Pp. 159—175 in Principles of Neural Science, E. R. Kandel and J. H. Schwartz, eds. New York: Elsevier.

Schwartz, J., C. Angle, and H. Pitcher. 1986. Relationship between childhood blood lead levels and stature. Pediatrics 77:281—286.

Schwartz, J., P. J. Landrigan, R. G. Feldman, E. K. Silbergeld, E. L. Baker, and I. H. Von Lindern. 1988. Threshold effect in lead-induced peripheral neuropathy. J. Pediatr. 112:12—17.

Seil, F. J., P. W. Lampert, and I. Klatzo. 1969. Neurofibrillary spheroids induced by aluminum phosphate in dorsal root ganglia neurons *in vitro*. J. Neuropathol. Exp. Neurol. 28:74—85.

Seppalainen, A. M. H., and M. Haltia. 1980. Carbon disulfide. Pp. 356—373 in Experimental and Clinical Neurotoxicology, P. S. Spencer and H. H. Schaumburg, eds.

Baltimore: Williams & Wilkins.

Sette, W. F. 1987. Complexity of neurotoxicological assessment. Neurotoxicol. Teratol. 9:411—416.

Shalat, S. L., B. Seltzer, and E. L. Baker, Jr. 1988. Occupational risk factors and Alzheimer's disease: A case-control study. J. Occup. Med. 12:934—6.

Shepard, T. H. 1989. A Catalog of Teratogenic Agents, 6th ed. Baltimore, Md.: Johns Hopkins University Press.

Shiraishi, S., N. Inoue, Y. Murai, A. Onishi, and S. Noda. 1983. Dipterex (trichlorfon) poisoning: Clinical and pathological studies in human and monkeys. Sangyo Ika Daigaku Zasshi. 5(Suppl.):125—132.

Silbergeld, E. K. 1986. Maternally mediated exposure of the fetus: In utero exposure to lead and other toxins. Neurotoxicology 7:557—568.

Silbergeld, E. 1990. Developing formal risk assessment methods for neurotoxicants: An evaluation of the state of the art. Pp. 133—148 in Advances in Neurobehavioral Toxicology: Applications in Environmental and Occupational Health, B. L. Johnson, W. K. Anger, A. Durao, and C. Xintaras, eds. Chelsea, Mich.: Lewis Publishers, Inc.

Silbergeld, E. K., and J. J. Chisolm. 1976. Lead poisoning: Altered urinary catecholamine metabolites as indicators of intoxication in mice and children. Science 192:153—155.

Silbergeld, E. K., and R. E. Hruska. 1980. Neurochemical investigations of low level lead exposure. Pp. 135—157 in Low Level Lead Exposure: The Clinical Implications of Current Research, H. L. Needleman, ed. New York: Raven Press.

Smith, H. V., and J. M. K. Spaulding. 1959. Outbreak of paralysis in Morocco due to orthocresyl phosphate poisoning. Lancet 2:1019.

Smith, M. I., E. Elvove, P. J. Valaer, Jr., W. H. Frazier, and G. E. Mallory. 1930. Pharmacological and chemical studies of the cause of so-called ginger paralysis. U.

S. Public Health Reports 45:1703—1716.

Smith, M. A., L. D. Grant, A. I. Sors, eds. 1989. Lead Exposure and Child Development: An International Assessment. Dorcrecht, the Netherlands: Kluwer Academic Publishers.

Sobotka, T. J. 1986. The regulatory perspective of diet-behavior relationships. Nutr. Rev. 44:241—245.

Sokoloff, L. 1989. Circulation and energy metabolism of the brain. Pp. 565—590 in Basic Neurochemistry, 4th ed., G. Siegel, B. Agranoff, R. Albers, and P. Molinoff, eds. New York: Raven Press.

Spencer, P. S., and H. H. Schaumburg. 1977a. Ultrastructural studies of the dying-back process. III. The evolution of experimental peripheral giant axonal degeneration. J. Neuropathol. Exp. Neurol. 36:276—299.

Spencer, P. S., and H. H. Schaumburg. 1977b. Ultrastructural studies of the dying-back process. IV. Differential vulnerability of PNS and CNS fibers in experimental central-peripheral distal axonopathies. J. Neuropathol. Exp. Neurol. 36:300—320.

Spencer, P. S., and H. H. Schaumburg, eds. 1980. Experimental and Clinical Neurotoxicology. Baltimore: Williams & Wilkins.

Spencer P. S., H. H. Schaumburg, R. L. Raleigh, and C. J. Terhaar. 1975. Nervous system degeneration produced by the industrial solvent methyl n-butyl ketone. Arch. Neurol. 32:219—222.

Spencer, P. S., M. C. Bischoff, and H. H. Schaumburg. 1978. On the specific molecular configuration of neurotoxic aliphatic hexacarbon compounds causing central-peripheral distal axonopathy. Toxicol. Appl. Pharmacol. 44:17—28.

Spencer, P. S., M. I. Sabri, H. H. Schaumburg, and C. L. Moore. 1979. Does a defect of energy metabolism in the nerve fiber underlie axonal degeneration in polyneuropathies? Ann. Neurol. 5:501—507.

Spencer, P. S., M. C. Bischoff, and H. H. Schaumburg. 1980. Neuropathological methods for the detection of neurotoxic disease. Pp. 743—757 in Experimental and Clinical Neurotoxicology, P. S. Spencer and H. H. Schaumburg, eds. Baltimore, Md.: Williams & Wilkins.

Spencer, P. S., J. Arezzo, and H. Schaumburg. 1985. Chemicals causing disease of neurons and their processes. Pp. 1—14 in Neurotoxicity of Industrial and Commercial Chemicals, Vol. 1, J. L. O'Donoghue, ed. Boca Raton, Fla.: CRC Press.

Spencer, P. S., P. B. Nunn, J. Hugon, A. C. Ludolph, S. M. Ross, D. N. Roy, and R. C. Robertson. 1987. Guam amyotrophic lateral sclerosis-parkinsonism-dementia linked to a plant excitant neurotoxin. Science 237:517—522.

Spyker, J. M. 1975. Assessing the impact of low level chemicals on development: Behavioral and latent effects. Fed. Proc. 34:1835—1844.

Srivastava, A. K., M. Das, and S. K. Khanna. 1990. An outbreak of tricresyl phosphate poisoning in Calcutta, India. F. Chem. Toxic. 28:303—304.

St. Clair, M. B. Genter, V. Amarnath, M. A. Moody, D. C. Anthony, C. W. Anderson, and D. G. Graham. 1988. Pyrrole oxidation and protein crosslinking as necessary steps in the development of γ-diketone neuropathy. Chem. Res. Toxicol. 1:179—185.

St. Clair, M. B. Genter, D. C. Anthony, C. J. Wikstrand, and D. G. Graham. 1989. Neurofilament protein in γ-diketone neuropathy: In vitro and in vivo studies using the seaworm Myxicola infundibulum. Neurotoxicology 10:743—756

Stark, M., T. Bilzer, N. Inoue, and W. Wechsler. 1989. Influence of 5-azacytidine on differentiation and growth in rat nervous system tumor cell lines. Neurotoxicol. Teratol. 11:557—561.

Stebbins, W. C., D. B. Moody, J. E. Hawkins, Jr., L. G. Johnsson, and M. A. Norat. 1987. The species-specific nature of the ototoxicity of dihydrostreptomycin in the patas monkey. Neurotoxicology 8:33—44.

Sterman, A. B., and H. H. Schaumburg. 1980. Neurotoxicity of selected drugs. Pp. 593—612 in Experimental and Clinical Neurotoxicology, P. S. Spencer and H. H. Schaumburg, eds. Baltimore, Md.: Williams & Wilkins.

Streissguth, A. P., H. M. Barr, P. D. Sampson, B. L. Darby, and D. C. Martin. 1989. IQ at age 4 in relation to maternal alcohol use and smoking during pregnancy. Dev. Psychol. 25:3—11.

Stryer, L. 1983. Transducin and the cyclic GMP phosphodiesterase: Amplifier protein in vision. Cold Spring Harbor Symposium. Quant. Biol. 48, Pt. 2:841—852.

Swygert, L. A., E. F. Maes, L. E. Sewell, L. Miller, H. Falk, and E. M. Kilbourne. 1990. Eosinophilia-myalgia syndrome. Results of a national surveillance. JAMA 264:1698—1703.

Szakál-Quin, G., D. G. Graham, D. S. Millington, D. A. Maltby, and A. T. McPhail. 1986. Stereoisomer effects on the Paul-Knorr synthesis of pyrroles. J. Org. Chem. 51:621—624.

Takahashi, M., T. Ohara, and K. Hashimoto. 1971. Electrophysiological study of nerve injuries in workers handling acrylamide. Int. Arch. Arbeitsmed. 28:1—11.

Takeuchi, T. 1977. Pathology of fetal Minamata disease: Effect of methylmercury on human intrauterine life. Paediatrician 6:69—87.

Teitelbaum, J. S., R. J. Zatorre, S. Carpenter, D. Gendron, A. C. Evans, A. Gjedde, and N. R. Cashman. 1990. Neurologic sequelae of domoic acid intoxication due to the ingestion of contaminated mussels. N. Engl. J. Med. 322:1781—1787.

Tilson, H. A. 1987. Behavioral indices of neurotoxicity: What can be measured? Neurotoxicol. Teratol. 9:427—443.

Tilson, H. A. 1990a. Animal neurobehavioral test battery in NTP assessment.

Pp. 403—418 in Advances in Neurobehavioral Toxicology: Applications in Environmental and Occupational Health, B. L. Johnson, W. K. Anger, A. DuRao, and C. Xintaras, eds. Chelsea, Mich.: Lewis Publishers, Inc.

Tilson, H. A. 1990b. Neurotoxicology in the 1990s. Neurotoxicol. Teratol. 12:293—300.

Tilson, H. A., and C. L. Mitchell. 1984. Neurobehavioral techniques to assess the effects of chemicals on the nervous system. Annu. Rev. Pharmacol. Toxicol. 24:425—450.

Towfighi, J., N. K. Gonatas, and L. McCree. 1974. Hexachlorophene neuropathy in rats. Lab. Invest. 31:712—21.

U.S. House of Representatives, Committee on Science and Technology. 99th Congress, Second Session. 1986. Neurotoxins: At Home and the Workplace. Report 99-827. Washington, D.C.: U.S. Government Printing Office.

Vasilescu, C. 1976. Sensory and motor conduction in chronic carbon disulphide poisoning. Eur. Neurol. 14:447—457.

Verity, M. A., T. S. Sarafian, W. Guerra, A. Ettinger, and J. Sharp. 1990. Ionic modulation of triethyllead neurotoxicity in cerebellar granule cell culture. Neurotoxicol. 11:415—426.

Veronesi, B., E. R. Peterson, G. DiVincenzo, and P. S. Spencer. 1978. A tissue culture model of distal (dying-back) axonopathy: Its use in determining primary neurotoxic hexacarbon compounds. J. Neuropath. Exp. Neurol. 33:703(Abs).

Vigliani, E. C. 1954. Carbon disulfide poisoning in viscose rayon workers. Br. J. Industr. Med. 11:325—344.

Vorhees, C. V. 1986. Comparison and critique of government regulations for behavioral teratology. In Handbook of Behavioral Teratology, E. P. Riley and C. V. Vorhees, eds. New York: Plenum Press.

Wagener, Diane K., and P. A. Buffler. 1989. Geographic distribution of deaths due to sentinel health event (occupational) causes. Am. J. Indust. Med. 16:355—372.

Waldron, H. A. 1973. Lead poisoning in the ancient world. Med. Hist. (London) 17:391—399.

Walker, C. H., W. O. Faustman, S. C. Fowler, and D. B. Kazar. 1981. A multivariate analysis of some operant variables used in behavioral pharmacology. Psychopharmacology (Berlin) 74:182—186.

Wallace, R. B., C. E. Daniels, J. Altman. 1972. Behavioral effects of neonatal irradiation of the cerebellum. 3. Qualitative observations in aged rats. Dev. Psychobiol. 5:35—41.

Weiss, B. 1990. Risk assessment: The insidious nature of neurotoxicity and the aging brain. Neurotoxicology 11:305—313.

Weiss, B., and T. W. Clarkson. 1986. Toxic chemical disasters and the implications of Bhopal for technology transfer. Milbank Q. 64:216—240.

West, J. R. 1990. Reply to: Response to letters dealing with warning labels on alcoholic beverages. Teratology 41:485—487.

WHO (World Health Organization). 1986. Principles and Methods for the Assessment of Neurotoxicity Associated with Exposure to Chemicals. Environmental Health Criteria Document 60. Geneva: World Health Organization.

Williamson, A. M. 1990. The development of a neurobehavioral test battery for use in hazard evaluations in occupational settings. Neurotoxicol. Teratol. 12:509—514.

Williamson, A. M., and R. K. C. Teo. 1986. Neurobehavioral effects of occupational exposure to lead. Brit. J. Indus. Med. 43:374-380.

Williamson, A. M., R. K. C. Teo, and J. Sanderson. 1982. Occupational mercury exposure and its consequences for behavior. Int. Arch. Occup. Environ. Health. 50:273—286.

Willis, W. D., Jr., and R. G. Grossman. 1973. P. 62 in Medical Neurobiology: Neuroanatomical and Neurophysiological

Principles Basic to Clinical Neuroscience. St. Louis: C. V. Mosby.

Winder, C., and I. Kitchen. 1984. Lead neurotoxicity: A review of the biochemical, neurochemical, and drug-induced behavioural evidence. Prog. Neurobiol. 22:59—87.

Wouters, W., and J. van den Bercken. 1978. Action of pyrethroids. Gen. Pharmacol. 9:387—398.

Wyzga, R. 1990. Towards quantitative risk assessment for neurotoxicity. Neurotoxicology 11:199—208.

Yamamoto, D., F. N. Quandt, and T. Narahashi. 1983. Modification of single sodium channels by the insecticide tetramethrin. Brain Res. 274:344—349.

Yamamura, Y. 1969. *n*-Hexane polyneuropathy. Folia Psychiatr. Neurol. Jpn. 23:45—57.

Yokel, R. A. 1983. Repeated systemic aluminum exposure effects on classical-conditioning of the rabbit. Neurobehav. Toxicol. Teratol. 5:41—46.

Yokel, R. A., S. D. Provan, J. J. Meyer, and S. R. Campbell. 1988. Aluminum intoxication and the victim of Alzheimer's disease: Similarities and differences. Neurotoxiciology 9:429—442.

Zec, R. F., and D. R. Weinberger. 1986. Relationship between CT scan findings and neuropsychological performance in chronic schizophrenia. Psychiat. Clin. North Am. 9:49—61.

Zeisler, R., S. H. Harrison, and S. A. Wise, eds. 1983. The Pilot National Environmental Specimen Bank: Analysis of Human Liver. National Bureau of Standards Special Publication 656. National Bureau of Standards, U.S. Department of Commerce, Washington, D.C. 128 pp.

Index

A

3-Acetyl-2,5-hexandedione (AcHD), 38
Acetylcholinesterase, 47, 48, 79
Acrylamide, 76, 80, 98, 101, 116, 119
Agency for Toxic Substances and Disease Registry (ATSDR), 1, 108-109
Alcohol, *see* Ethanol
Alzheimer's disease, 15, 40, 47, 48, 51, 83
American Conference of Governmental Industrial Hygienists (ACGIH), 17, 53
Ames test, 59, 112
γ-Aminobutyric acid (GABA), 35, 63, 79, 117, 118
Amyotrophic lateral sclerosis (ALS), 15, 40, 51
Animals
  FOB end points, list, 71
  models, 49, 54, 79
  neurotoxicity testing, 49-50, 53-93, 118, 119, 124, 125
  neurotoxic effects of representative agents, list, 67-68
  see also Testing
Annual Survey of Illness and Injury, 106
Antidepressants, tricyclic, 16
Australian information-processing theory test battery, 97, 102, 103
Axons, *see under* Neurons

B

Behavioral assessment, 66, 69-70, 72-73, 87
Benchmark dose (BD), 114
Biochemical assays, 59-62
Biologic markers,
  animal models, 49, 54
  disease surveillance, 108,
  concepts and defintions, 43-48, 51-52
  of effect, 44, 45, 47-48, 51
  of exposure, 16, 36, 43, 44, 45, 46-47, 51
  for in vitro systems, list, 63
  of neurotoxicity, list, 45
  quality assurance and quality control, 49-50
  recommendations, 5, 124
  selected, illustration, 92
  of susceptibility, 44, 45, 48, 51
  use in risk assessment, 43, 50-51, 124
  validation, 44, 48-50, 56, 124
Blood-brain barrier, 16, 28-29, 39, 55, 64
Brain, imaging procedures, 81-83
Bureau of Labor Statistics (BLS), 106

C

Carbamates, 58-59
Carbon disulfide, 11

149

Cells
    culture techniques, 61-62, 63-64
    ganglia, 27-28
    glial, 21, 22, 24, 39, 61, 62, 63, 64, 79
    loss, 51
    nuclei, 27, 28
    plasticity, 29, 30
    Schwann, 22, 24
    see also Nervous system; Neurons
Centers for Disease Control, 50
Central nervous system (CNS), 27-29, 30, 32,
    34, 35, 40, 47, 55, 64, 80, 81, 82, 84, 97,
    103, 111, 115-116, 120
    see also Nervous system
Cerebral cortex, 27-28
Cerebral glucose, 83
Chemicals
    diagram of MPTP toxicity, 39
    exposure response characteristics, list, 98-
        99
    neurobehavioral effects, list, 11, 69-70
    neurotoxic effects in humans and animals,
        list, 67-68
    neurotoxicants, list, 10
    receptors, list, 118
    regulation and testing, 18-19, 53-93, 95,
        97, 123-127
    see also Exposure; Nervous system; Risk
        assessment; Testing; and specific chem-
        icals
Comprehensive Environmental response,
    Compensation, and Liability Act
    (CERCLA), 109
Computed axial tomography (CAT), 31, 81,
    82, 83, 109-110
Consumer Product Safety Commission, 86,
    124
Cycad (Cycas circinalis), 15

D

DDT, 25, 31, 34
Department of Veterans Affairs, 126
Developmental neurotoxicology, 84
Dieldrin, 117
Diisopropyl fluorophosphate (DFP), 75

γ-Diketones, 31, 32-33, 36-39, 40, 46, 48, 59,
    64, 77, 79
3,4-Dimethyl-2,5-hexanedione (DMHD), 37,
    38
Disease surveillance
    biologic markers, 108
    clinical neurologic examination, list of
        components, 100
    disease and exposure registries, 108-109
    identification and prevention efforts, 95-
        97, 120
    imaging techniques, 109-110
    medical training, 107
    recommendations, 126-127
    sentinel health events (SHEs), 108, 127
    standardized disease definitions, 107-108
    surveillance efforts, 105-110
Dopamine, 29, 31, 33, 39, 40, 47, 78-79
Dose-response relationships, 2-3, 5, 6, 16, 43,
    50, 51, 56, 72, 111, 112, 113, 114-116, 120,
    125
Dyskenesia, 15, 54

E

Electroencephalography (EEG), 78
Electrophysiology, 31
Environmental Protection Agency (EPA)
    guidelines, 84, 87
    premanufacturing notice (PMN) program,
        86
    regulation and testing of chemicals, 18,
        19, 50, 53-54, 58, 86-87, 89, 105, 124
    review of substances, 53, 86-87
Epinephrine, 48
Ethanol, 16, 17-18, 98
Explant cultures, 61, 62
Exposure
    biologic markers, 16, 36, 43, 44, 45, 46-47,
        51, 126
    characteristics of neurotoxicant exposure
        responses, list, 98-99
    detection and control, 18-19, 43, 95-110
    environmental, 1-2, 16-19, 95, 96
    magnitude of problem, 17-18
    occupational, 2, 9, 16, 17, 96

during pregnancy, 15, 50-51
see also Risk assessment

F

Federal Insecticide, Fungicide, and Rodenticide Act (FIFRA), 86, 87
Finland Institute of Occupational Health test
battery, 97, 101, 102
Food and Drug Administration (FDA), 50,
58, 84, 87, 124
Fuctional observational batteries (FOBs), 66-
72, 73, 74, 89, 90

G

Glial fibrillary acidic protein (GFAP), 79
Glucose, in nervous system, 32
Good laboratory practices (GLPs), 50

H

Hazard identification, in neurotoxicity risk
assessment, 4, 112, 120, 125-126
Health Care Financing Administration (U.S.
Department of Health and Human Services), 18
Health Hazard Evaluation Program, 97, 104,
107
Hemoglobin adducts, 79
n-Hexane, 33-34, 36, 37, 38, 59, 77, 98, 101,
121
2,5-Hexanedione (HD), 36, 37, 38, 47, 80
Homovanillic acid (HVA), 78
Hormone production, 80
Humans
effects of representative neurotoxic
agents, 67-68
neurotoxicity testing, 95-110, 118-119, 124,
125
see also Risk Assessment; Screening;
Surveillance; Testing
Hypothalmus, 27-28

I

Imaging procedures, 81-83, 109-110
In vitro and in vivo testing, *see* Screening;
Testing
Insecticides, 25, 34-36, 44, 59
see also specific insecticides
Integrated Management Information Systems
(IMIS), 104, 105
Integrative functions
cognitive processes, 66, 70
motor performance, 65, 69, 72, 73
neurotoxic effects, 65-66, 69-70
sensory acuity, 65-66, 69, 72, 73-74
memory, 66, 70, 74-76
Intelligence quotient (IQ), 9, 64, 112, 13

L

Lathyrism (spastic paraparesis), 11
Lead, 1, 2, 11, 15, 55, 58, 72, 78-79, 80-81,
98, 105, 108, 113, 116, 121
Learning/memory function, 66, 70, 74-76
Lowest-observed-adverse-effect level
(LOAEL), 56, 114-116
Lymphocyte neurotoxic esterase (NTE), 79
Lysergic acid diethylamide (LSD), 33

M

Magnetic resonance imaging (MRI), 31, 81-
82, 83, 110
Malathions, 31
Manganese, 15
Mental retardation, 15
Mercury, 1, 2, 15, 51, 58, 79, 98
Metabolism, in central nervous system, 28
Methanol, 1, 120
Methyl *n*-butylketone, 36, 37, 48, 59, 77, 101
1-Methyl-4-phenyl-1,2,3,6-tetrahydropyridine,
see MPTP
Methylmercury, 49, 51, 55, 64, 72, 80, 85, 98,
101, 120
Microencephaly, 15
Modifying variables, in neurotoxicity testing

age at exposure, 83-84
age at testing, 84-85
genetic differences, 85
sex, 85
Motor activity and function, 65, 69, 72, 73, 77
MPTP (1-methyl-4-phenyl-1,2,3,6-tetrahy-
dropyridine), 15, 31, 34, 39-40, 49, 51, 58,
82-83, 98, 110, 119
Myelin, 24, 30, 31, 38, 64, 80-81

N

National Center for Health Statistics
(NCHS), 106, 108, 126
National Center for Toxicological Research
(NCTR)
behavioral-teratology tests, list, 83
Collaborative Study, 84
National Death Index, 106
National Health and Nutrition Examination
Surveys (NHANES), 105, 106
National Health Interview Survey, 106
National Human Adipose Tissue Survey
(NHATS), 105
National Institute for Occupational Safety
and Health (NIOSH), 17, 97, 104-105,
107, 108
National Institute of Standards and Tech-
nology, 50
National Occupational Exposure Survey
(NOES), 104-105, 109
National Occupational Hazard Survey
(NOHS), 104-105, 109
National Toxicology Program (NTP), 87-88
Nerve-conduction studies, 76-77
Nervous system
biologic markers, 43-52
cellular anatomy and physiology, 15-16,
21-27, 61, 62, 63-64
complexity, 29-30, 31
effects of toxicants on, 2, 9-17, 21, 25, 30-
40
examples of neurotoxic mechanisms, 33-
40
lipid-soluble toxicants, 31-32
neuropathic evaluation areas, list, 81
receptor sites, 46, 47
repair limitations, 31, 40, 51, 55

structure and function, 15-16, 21, 27-30,
40-41
tissues used in neurophatologic evaluation, list, 82
see also Central Nervous System; Neurons; Neurotransmitters; Peripheral
Nervous System
Neurobehavioral Core Test Battery (NCTB),
97-101, 102, 103-104
Neurobehavioral Evaluation System (NES),
101, 102, 104
Neurobehavioral test batteries, 97-104
see also specific test batteries
Neurochemical procedures, 78-79
Neuroendocrine interactions, 79-80, 125
Neurons
axons, 22-24, 25, 30, 31, 32-33, 38, 55, 64,
77, 80
communication, 24-27, 33, 55
development and functions, 21-22, 29-30,
51
glucose, 32
illustrations, 23, 26
integrative functions, 65-66
ions, 24-25, 29, 33, 116
nerve impulse, 24-25, 26
structure, 22-24
synaptic messengers, 26-27, 33, 55
synaptic transmission, 25-26, 27, 33, 77
trophic interactions, 30
see also Nervous system; Neurotransmitters
Neuropathology, 80-81
Neurophysiologic procedures, 76
Neurotoxicity events, list of major, 12-14
Neurotransmitters, 25-27, 31, 32, 33, 40, 47,
55, 61, 78, 111-112, 116
see also Nervous system; Neurons
No-observed-adverse-effect level (NOAEL),
56, 114-116
Nuclear magnetic resonance spectroscopy
(MRS), 81, 82

O

Occupational Safety and Health Administration (OSHA), 104, 105, 109, 124
Office of Technology Assessment (OTA), 90

Office of Toxic Substances (EPA), 87
Organization for Economic Cooperation and Development (OECD), 50, 86, 90
Organ cultures, 61-62
Organophosphates, 47, 58-59, 77, 98
Organophosphorus (OP) esters, 59-60, 79, 81, 86

P

Parkinsonism, 15, 31, 34, 39-40, 49, 51, 85, 119
Peripheral nervous system, 27, 31, 35, 81, 116
  see also Nervous system
Pesticides, 1, 2, 17, 31-32, 33, 34, 47, 72, 79, 86, 95, 105, 117
  see also specific pesticides
Pharmaceuticals, 2, 9, 16, 18, 33, 54
Phenylalanine, 85
Poisoning, pyrethroid, 34-36
Positron-emission tomography (PET), 31, 40, 81, 82, 83, 110
Potassium ions, 24, 25, 33
Pregnancy, 15, 50-51
Prevention (primary and secondary), 18, 53, 95-96
Psychosis, toxic, 15
Pyrethroids, 25, 31, 33, 34-36, 44, 59
Pyrrole, 37-38, 59

R

Radiation, 9, 15
Reaggregate cultures, 61, 62
Risk assessment
  approaches to, 114-116
  biologic markers, 43, 50-51, 124, 124, 125
  cognitive model, 113
  curve-fitting, 114, 116-117, 120
  carcinogenicity models, 111-113, 119
  hockey-stick model, 116-117
  neurotoxicity models, 111-113, 117-119, 120
  neurotoxicity receptors, 117-118

no-threshold model, 117-118
NRC paradigm, 56
problems, 17-18, 19, 50-51, 111-112
recommendations, 5-6, 123-127
safety-factor approach, 114-117
statistical models, 114-116, 121
strategy, 2-3
  see also Screening; Surveillance; Testing

S

Schedule-controlled operant behavior (SCOB), 72-73
Screening
  applications, 57-58
  evaluation of in vitro tests, 91-92
  in vitro batteries, 62-65, 90-92
  in vivo batteries, 66-71
  multitiered testing, 2-3, 88-89, 124-125
  priority-setting and implementation, 90
  protocol for in vitro system, 65
  research needs, 2-3, 90-93
  sensitivity, 56-57
  specificity, 56-57
  strategies for improved testing, 88-89, 124-125
  validation, 3, 57, 89-90, 125
  see also Testing
Sensory-evoked potentials, 77-78
Sensory function, 65-66, 69, 72, 73-74, 77-78
Sentinel Event Notification System (SENSOR), 108
Social Security Administration, 126
Sodium ions, 24-25, 33, 34, 35-36, 44, 59, 63
Stimulus-response techniques, 74, 75-76
Structure-activity relationships (SARs), 4, 19, 54, 56, 58-59, 117, 124, 125
Superfund Amendments and Reauthorization Act (SARA), 105
Surveillance
  of disease, 105-110
  of exposure, 104-105
  need, 95-97, 126
  neurobehavioral test batteries, 97-104
  recommendations, 4-5, 126-127
  see also Disease surveillance; Risk Assessment

T

Testing
  behavioral assessment, 66
  diagnostic techniques, 31, 47-48
  difficulties, 54-55
  effects of age, 83-85
  effects of genetic differecnes, 85
  effects of sex, 85
  electroencephalography, 78
  functional observational batteries (FOBs),
    66-72, 73, 74, 89, 90
  hazard characterization, 56
  imaging procedures, 81-83
  in vitro, 4, 5, 18-19, 54, 59-65, 90-92, 124,
    125
  in vivo, 5, 18-19, 54, 65-85, 124, 125
  laboratory, 49-50; *see also* Animals
  modifying variables, 83
  motor-activity, 72
  nerve-conduction studies, 76-77
  neurochemical procedures, 78-79
  neuroendocrine interactions, 79-80, 125
  neuropathology, 80-81
  neurophysiologic procedures, 76
  recommendations, 2-5
  regulatory approaches, 86-93
  schedule-controlled operant behavior, 72-
    73
  screening characteristics and applications,
    56-58
  sensory evoked potentials, 77-78
  specialized function tests, 73-76
  standards and guidelines, 18, 19, 49-50,
    53-54
  strategy, 2-4, 56, 125
  structure-activity relationships, 19, 54, 56,
    58-59
  see also Animals; Environmental Protec-
    tion Agency; Humans; Screening; Sur-
    veillance
Toxic Release Inventory (TRI), 105
Toxic Substances Control Act (TSCA), 19,
  86, 95
Tri-*o*-cresylphosphate (TOCP), 15

W

World Health Organization (WHO)
  recommendations for neuropathologic
    evaluation, 81
  Neurobehavioral Core Test Battery, 97-
    101, 102, 103-104